Reliability
and
Maintainability
Guideline
for
Manufacturing Machinery
and Equipment

Published by:

Society of Automotive Engineers, Inc.
400 Commonwealth Drive
Warrendale, PA 15096-0001
U.S.A.
Phone: (412) 776-4841
Fax: (412) 776-5760

and

National Center for Manufacturing Sciences, Inc.
900 Victors Way
Ann Arbor, MI 48108
U.S.A.
Phone: (313) 995-0300

Library of Congress Catalog Card Number: 92-63367
ISBN 1-56091-362-2

SAE Order No. M-110

Acknowledgement

A publication of this size and scope only goes together with the cooperation of many people and organizations.

We wish to acknowledge the pioneering efforts of Ford Motor Company who published the first guideline in 1990 and have aggressively implemented it in a number of operating units within their organization.

General Motors Corporation who, with a guideline of their own, had the vision that a common guideline would be of great benefit in broadening the practice of Reliability & Maintainability (R&M) among the parts manufacturing community.

Chrysler Corporation who, with their TESQA specifications provided some of the earliest application of reliability engineering practice to machine building.

We acknowledge and thank members of the Inner Core Group who spent many hours studying the intricacies of the practice of R&M and resolved the wording, structure and organization of this Guideline. In addition to the above, the membership includes: Cincinnati Milacron Inc., Giddings and Lewis, Ingersoll Milling Machine Co., Ingersoll-Rand, Litton-Landis, PICO, The Boeing Company, Pratt & Whitney, AT&T, and Eastman Kodak Company.

With appreciation we also acknowledge the important contributions made by other organizations through individual meeting deliberations, their support and contributions to the Reliability and Maintainability Guideline for Manufacturing Machinery and Equipment, and discussions at the Advisory Group meetings. These include: The Association For Manufacturing Technology (AMT), Allen-Bradley, Caterpillar, Cummins Engine Company, Inc., Deere and Co., Digital Equipment Corporation (DEC), Lamb-Technicon, Midwest Brake Bond Co., Rockwell International, Saginaw Machine Systems (SMS), Texas Instruments (TI), Society of Automotive Engineers (SAE), and Kingsbury Corporation.

Lastly, we acknowledge the efforts of three organizations who brought these companies together and provided the forum and environment in which they could work together for the common good. They are The National Center for Manufacturing Sciences (NCMS), The Industrial Technology Institute (ITI), and Western Michigan University.

EXECUTIVE OVERVIEW

This guideline consolidates Reliability and Maintainability (R&M) terminology, methodology and procurement language generally accepted by suppliers and users of equipment employed for the manufacture of discrete components. This composite framework will help integrate R&M concepts when the equipment is designed, and contribute to the reduction of maintenance, warranty, and life cycle costs while increasing equipment availability.

R&M is a discipline. It is steeped in well documented techniques that are meant to direct both machine suppliers and users beyond the question of "Will it work?" to a quantifiable analysis of how long it will work without failure.

The requirements for R&M encourage a partnership between the supplier and user of manufacturing machinery and equipment. Both members of this partnership must understand what equipment performance data is needed to ensure continued improvement in equipment operation and design, and must exchange this information on a regular basis. The successful implementation of a R&M program requires a strong commitment from both the user and supplier management teams.

The R&M process developed for manufacturing machinery and equipment has been organized into a five-phase program management process that includes Concept, Development/Design, Build and Install, Operation and Support, and Decommissioning/Conversion, with accompanying implications for suppliers and users.

Application-specific R&M techniques can be prescribed for each unique equipment acquisition by employing standardized worksheets that are keyed to an appropriate R&M matrix. The process is further supported by appendices that enhance the understanding of the underlying technical aspects of R&M. In the longer run this guideline will contribute to higher levels of productivity and competitiveness, and will be subject to change and modification as dictated by the industry's aspiration to be at the leading edge.

This guideline will remain a dynamic document, being expanded and refined consistent with the requirements of the manufacturing machinery and equipment marketplace. Standardization of R&M engineering principles can avoid costly duplication of efforts and enhance their competitive positions in the marketplace.

The majority of this work was done with personnel from transportation-related industries, and the guideline reflects this orientation.

Table of Contents

Table of Contents - Continued

Table of Contents - Continued

Mission

This guideline is intended to provide a description of reliability and maintainability (R&M) fundamentals for manufacturing machinery and equipment user and supplier personnel at all operating levels. It embraces a concept of up-front engineering and continuous improvement in the design process for machinery and equipment. The document is not intended to be a primer on R&M. It simply presents standard techniques as they apply to the life cycle of machinery and equipment, and gives a sequence of R&M actions to be followed.

SECTION 1

INTRODUCTION AND BENEFITS

Reliability and Maintainability are key ingredients to preserving production efficiency and lead to lower total life cycle costs that are necessary to assert a competitive edge.

INTRODUCTION

- **Reliability** is the probability that machinery/equipment can perform continuously, without failure, for a specified interval of time when operating under stated conditions. Increased reliability implies less failure of the machinery and consequently less downtime and loss of production.

- **Maintainability** is a characteristic of design, installation and operation, usually expressed as the probability that a machine can be retained in, or restored to, specified operable condition within a specified interval of time when maintenance is performed in accordance with prescribed procedures.[1]

Reliability and Maintainability (R&M) are vital characteristics of manufacturing machinery and equipment that enable U.S. manufacturers to be world class competitors. Efficient production planning depends on a process that yields high quality parts at a specific rate without interruption. Predictable R&M of the manufacturing machinery and equipment is a key ingredient in maintaining production efficiency and the effective deployment of "Just-In-Time" principles. Improved R&M leads to lower total life cycle costs that are necessary to maintain the competitive edge.

This document supports these objectives by providing R&M techniques, and guidance on where to apply them, both in the up-front design and development of new equipment and in continuous improvement of the machinery after installation. Successful implementation of this R&M Guideline requires a cooperative effort between user and supplier. Neither participant in the process can accomplish the objectives alone.

---

[1]Other terms used in this guideline are defined in the glossary.

BENEFITS OF R&M

Improved R&M results in improved availability. Highly available production machinery offers the means for producing consistently high quality products at lower costs and at higher output levels. Successful application of R&M techniques has a very positive effect on employee morale and pride since reducing downtime also significantly reduces employee stress and frustration.

Table 1-1 shows how improved equipment R&M benefits both the user and the supplier.

Table 1-1. R&M User/Supplier Benefits	
User Benefits	**Supplier Benefits**
• Higher machinery and equipment availability	• Reduced warranty costs
• Unscheduled downtime reduced/eliminated	• Reduced build costs
• Reduced maintenance costs	• Reduced design costs
• Stabilized work schedule	• Improved user relations
• Improved J-I-T performance capability	• Higher user satisfaction
• Improved profitability	• Improved status in the marketplace
• Increased employee satisfaction	• A competitive edge in the marketplace
• Lower overall cost of production	• Increased employee satisfaction
• More consistent part/product quality	• Increased understanding of product applications
• Less need for in-process inventory to cover downtime	• Increased sales volume
• Lower equipment LCC	

Safety

Throughout any R&M program, consideration must always be given to safety. The benefits of an improved design must not be allowed to compromise the ability of the manufacturing machinery and equipment to be operated safely, and to be maintained without risk to personnel. Passive safety features should always be a criteria for good design.

Reduced Life Cycle Cost

Life Cycle Cost (LCC) refers to the total cost of a system during its life cycle (See Glossary definition of LCC and Appendix C). LCC is the sum of non-recurring costs plus operation and support costs. Figure 1-1 shows that operation and support typically consume about 50% of the total LCC (see Bibliography, reference 1).

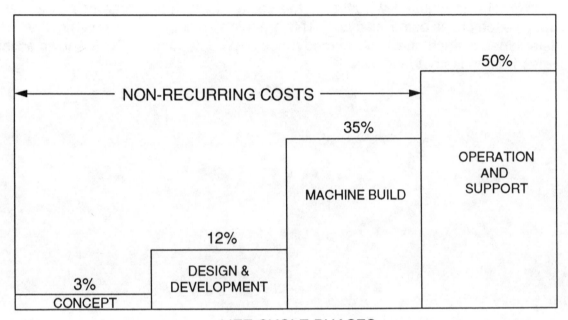

Figure 1-1. Total Life Cycle Cost

The total LCC can be lowered by emphasizing R&M during the conception and development stages. By using R&M to minimize stress (electrical, mechanical, thermal, etc.), the equipment will be less prone to failure during operation and the operation and support costs that account for the bulk of total LCC will decrease.

Figure 1-2 illustrates how a slight increase in spending to incorporate R&M practices during the concept and design stages can dramatically lower the operation and support costs. It is important to consider R&M at the early stage of a program. Industrial studies have shown that as much as 95% of LCC is determined during concept and development stages. Once new equipment has reached the build stage, therefore, only a 5% opportunity remains to improve the reliability or maintainability of the equipment.

The lower graphic in Figure 1-2 illustrates the points at which the three design reviews occur within each phase of the life cycle. This graphic also depicts how much of the LCC is predetermined at each phase. A shaded range of predetermined LCC is shown as the cost varies from industry to industry.

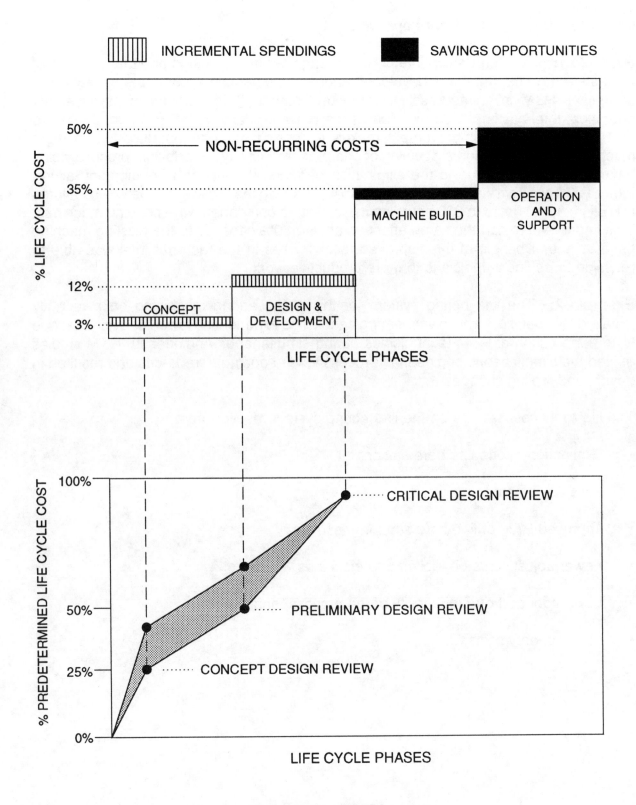

Figure 1-2. R&M Impact

Examples of Life Cycle Cost Improvement

Example 1: Intel Corporation is a U.S. firm engaged in the design and manufacture of solid-state devices. Intel has developed and is implementing a corporate strategy that addresses R&M in an aggressive, committed manner. In portions of its assembly operation, Intel has improved the Mean Time Between Failures (MTBF) from 5 minutes to 16 minutes. This improvement makes it possible for one operator to run eight machines rather than four, a doubling of operator productivity. In addition, process yields have been improved due to the elimination of scrap attributable to equipment failure. Intel's R&M program was also responsible for improving the Mean Time Between Failures (MTBF) from 10 hours to 250 hours on its solid-state component wire bonding machines. This improvement had the same effect as adding 30% capacity to the existing machine base. Another benefit of this improved reliability lies in the fact that Intel was able to reassign three line technicians to more productive work.

Example 2: The line boring system for the main bearing machining on a recently installed cylinder block and lower bearing/saddle line experienced lower life cycle cost due to the application of R&M technologies during simultaneous engineering. R&M studies guided the simultaneous engineering process toward optimum trade-offs and resulted in design and tooling changes.

The life cycle cost savings on the line boring system resulted from

- Elimination of one line bore station,

- Less equipment to maintain,

- Improved R&M of line bore support machinery,

- Fewer adjustments on tooling and machines,

- Lower tool cost due to stability of the process, and

- Lower energy costs.

SECTION 2

IMPLEMENTING R&M THROUGH THE LIFE CYCLE PROCESS

R&M should be implemented during all phases of the equipment life cycle.

INTRODUCTION

Successful implementation of R&M depends upon thorough communication between the user and the supplier. This communication must begin at project conception and continue through the entire life of the equipment to ensure that equipment problems will be identified, root causes determined, and corrective action implemented.

Attainment of reasonable levels of R&M seldom occurs by chance. It requires planning, goal definition, a design philosophy, analysis, assessment, and feedback for continuous improvement. Management must recognize the value of R&M and commit the manpower and resources necessary to attain the goal. Without such a commitment, the probability of attaining R&M goals is low. Successful attainment of the R&M quantitative and qualitative objectives requires a team effort involving all functions of the business. These objectives are detailed in this section as well as throughout these guidelines.

Figure 2-1, shows seven key practices for successful R&M implementation.

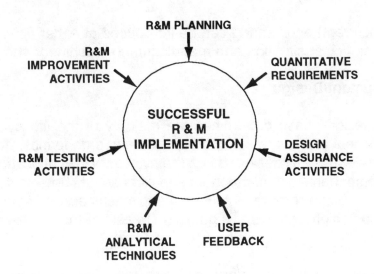

Figure 2-1. Keys to Successful Implementation

FIVE-PHASE PROGRAM MANAGEMENT PROCESS

Figure 2-2 surveys the typical product development process. It starts in Phase 1 with concept and proceeds through decommissioning and/or conversion in Phase 5. This process is appropriate for any hardware development program for machinery and equipment.

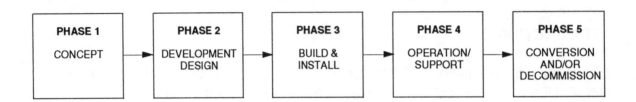

Figure 2-2. Five Phases of Manufacturing Machinery and Equipment Life Cycle

Phase 1 - Concept

The first phase is research and limited development or design, and usually results in a proposal. During this phase both the user and the supplier must work together to establish system requirements. The user team should include machine operators, maintenance personnel, and engineering personnel. The supplier team should include, engineering, service and assembly floor personnel, and sub-suppliers. Machinery mission and environmental requirements are defined during this phase, as are safety issues, goals for R&M, and goals for life cycle cost.

Simultaneous (concurrent) engineering can be introduced at either Phase 1 or Phase 2 depending on the particular situation and manufacturing machinery and equipment.

Phase 2 - Development/Design

The development/design phase determines the majority of the life cycle cost. All the issues from the concept phase are incorporated. Safety, ergonomics, accessibility, and other maintainability issues are designed into the system. R&M allocation requirements are formalized. Components and component suppliers should be selected based on the predictive R&M statistics they provide. Manufacturing machinery and equipment suppliers should utilize methods highlighted in this guideline to assure that R&M requirements will be met.

The design review ensures that the planned design is likely to meet all requirements in the most cost effective way, considering all variables and constraints and with special attention to maintainability. Section 4 amplifies design review procedures. Typically there are two design reviews in Phase 2. A preliminary review precedes commitment to a final design approach. A critical design review determines overall readiness for production prior to release of drawings to the manufacturing function.

Design review sessions should be held regularly to ensure clear communication between the user and manufacturing machinery and equipment suppliers. Operators, maintenance personnel, and product engineers should participate in the design review so that all concerned understand the design intent.

At this phase, the design must include suitable test plans, agreed to by both the user and the supplier, to demonstrate compliance to requirements, and must provide for teardown and reassembly in the user's plant. Responsibility for data collection, analysis and reporting is negotiated.

The supplier and the user should negotiate meaningful R&M goals and requirements for future monitoring and divide responsibility for collecting, analyzing and reporting of data.

Phase 3 - Build and Install

During the manufacturing and assembly of the machine, the achievement of R&M requirements should be monitored. Issues that could affect R&M must be communicated back to the design engineers to ensure that any redesign includes R&M improvements. Manufacturing process variables affecting R&M should be identified and targeted for control.

Several events that occur during Phase 3 require extra concern.

• Maintenance procedures are developed. A user representative should be involved in this process.

• Training starts at this phase and continues to the next phase.

• Machine acceptance testing, as agreed to in Phase 2 should be performed prior to teardown and installation.

• R&M data base collection begins during machine acceptance testing. All problems encountered during this phase should be documented for future reference and as candidates for continuous improvement.

- The machine will be transferred from the manufacturing machinery and equipment supplier's location to the user's plant. Critical assembly processes should be identified during teardown.

- Installation is a critical step. The machine has to be reassembled to the build requirements. Special attention should be given to the critical assembly processes identified during teardown.

- Some infant mortality failures may be present during initial startup. Every effort should be taken to eliminate infant mortality failures during the installation and debugging period.

Phase 4 - Operation & Support

In this phase, the equipment has been delivered and installed at the user location and is fully operational. Data collection and feedback are very important at this phase. Data collection mechanisms should be in place and agreed upon by both parties. Information collected during this phase is used to facilitate R&M growth and continuous improvement. During this phase preventive maintenance must be performed regularly. For a R&M initiative to be successful, the manufacturing machinery and equipment and component suppliers must have access to maintenance records and R&M data bases.

Phase 5 - Conversion and/or Decommissioning

This phase is the end of the expected life of the machine. If an increasing failure rate has resulted in increasingly expensive maintenance, the machine may require decommissioning, or may be rebuilt to a good-as-new state. Alternatively, the machine may still be in good condition but production needs may have changed, requiring the machine to go through major conversion to be used for production of other products. When either decommissioning, rebuilding, or conversion occurs, the feedback from the user plant should be recorded. The information will be used for R&M growth and continuous improvement in future generations of machinery.

This phase will not be addressed in the document, because no R&M activities are typically done in this phase.

SUMMARY OF PHASES

Figure 2-3 summarizes the five phases and what should be accomplished in each.

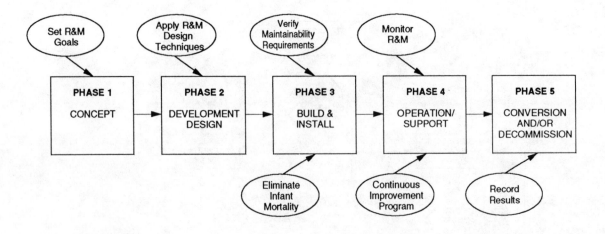

Figure 2-3. Goals of Each Phase

R&M APPLICATION GUIDE

R&M for machinery and equipment must be addressed in a systematic fashion. Figure 2-4 provides an example of the process flow for a R&M study.

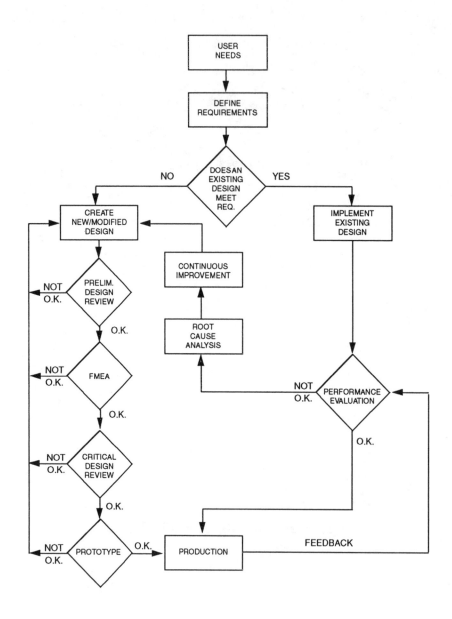

Figure 2-4. R&M Application

TAILORING R&M ACTIVITIES OVER THE LIFE CYCLE PHASES

R&M activities required during the equipment life cycle phases are a function of both hardware requirements and the capabilities of the selected supplier. Tailoring of these requirements by the user ensures a cost effective application of R&M program elements. Figure 2-5 provides an example of a generic R&M program matrix. The X's indicate when a R&M program activity is typically performed. For each equipment procurement, a similar Program Matrix will be developed to tailor the R&M needs of that procurement. Specific tailoring is a function of the amount of new development required, the end item application and the R&M capabilities of the selected supplier. Examples of this tailoring process can be found in Section 7, R&M and Contracting.

Generic R&M Program Matrix

	Concept – Phase 1	Design & Develop – Phase 2	Build & Install – Phase 3	O & S – Phase 4	Disposal – Phase 5
Reliability Requirements	X				
Maintainability Requirements	X				
Failure Definition	X				
Environment/Usage	X				
Design Margin		X			
Maintainability Design		X			
Reliability Predictions		X			
FMEA/FTA		X			
Design Reviews		X			
Parts			X		
Tolerance Studies			X		
Stress Analysis			X		
R Qualification Testing			X		
R Acceptance Testing			X		
R Growth/M Improvement				X	
Failure Reporting				X	
Data Feedback				X	

Figure 2-5. Sample R&M Requirements Worksheet

SECTION 3

USER AND SUPPLIER R&M ACTIVITIES IN THE CONCEPT PHASE (PHASE 1)

In the concept phase:

The machinery user specifies the reliability and maintainability requirements. The supplier should be prepared to address how the R&M qualitative and quantitative requirements dictated by the user will be met.

INTRODUCTION

During Phase 1 performance and R&M requirements are established, preliminary design concepts considered and proposals evaluated. This process leads to the selection of an equipment supplier. Simultaneous (concurrent) engineering can be introduced at either Phase 1 (concept) or the following Phase 2 (design and development) depending upon the particular situation and equipment involved.

USER RESPONSIBILITIES

Reliability Requirements

The machinery user specifies the reliability requirements in terms of

- Mean Time Between Failure (MTBF) or Mean Cycles Between Failure (MCBF) for repairable equipment, and

- Mean Time to Failure (MTTF) or Mean Cycles To Failure (MCTF) for non-repairable items.

Maintainability Requirements

The machinery user specifies maintainability requirements in terms of Mean Time To Repair or Mean Time To Replace (MTTR).

The values assigned to the R&M requirements indicate the inherent R&M designed into the machine. While the supplier will not be held accountable for nonconformance caused by operations outside stated conditions, both the supplier and user should recognize the impact of these real world occurrences. The definition of failure (see Glossary) considers all downtime events.

Machinery Use

All machinery suppliers and users should be aware of how the machinery is intended to be used and the impact of this use on reliability. Decreased cycle times obtained by increasing speeds and feeds have a negative impact on reliability. Similarly, increased complexity and extending the length of in-line operations drives up the failure rate. Minimizing, through design, the time required to perform scheduled maintenance in terms of accessibility, inspection, and service supports the intended availability of machines and equipment.

Duty Cycle

Many machinery components are sensitive to start-up and/or power-up. Therefore, anticipated duty cycles for machinery components should be understood.

Machinery Environment

Machinery specifiers and designers must fully understand the environments to which the machinery and its components will be subjected. These environmental elements include heat, humidity, contamination, shock, and vibration at various locations within and around the machine. These factors influence the performance of mechanical, hydraulic, pneumatic, electrical, and electronic components (including items contained in enclosures) to different degrees. Appendix A amplifies environmental considerations.

Continuous Improvement Monitoring

The suppliers can be greatly aided in the design of new manufacturing machinery and equipment by feedback from the user concerning any improvements made on the supplier's existing products currently in operation.

This feedback to the supplier is best achieved if the user supplies it up-front. Alternatively, the supplier must solicit input from the user's manufacturing engineer who, in turn, can coordinate gathering information from the plant's manufacturing operations. This information is provided by plant production and maintenance personnel in terms of

- The types of machine problems experienced,

- Corrective actions or improvements made, and

- The results of the improvements.

These improvements may be in the areas of operating cost, safety, ergonomics, or quality. In many cases the supplier is very familiar with these manufacturing machinery and equipment improvements on the user's floor due to involvement with the user plant's continuous improvement activity. This type of partnership provides more timely and effective results in incorporating improvements into the design of new manufacturing machinery and equipment.

Decisions to decommission, rebuild, or improve manufacturing machinery and equipment should involve all disciplines at the plant, to assure that the appropriate information-gathering and analysis is completed prior to disposition of the manufacturing machinery and equipment.

Life In Terms of Throughput

During the concept and development stage, the plant manufacturing engineer should provide the supplier with clear definitions of the throughput and product mix requirements over the projected life of the equipment. Clear definitions are necessary to ensure the supplier's understanding so that an efficient design-for-use approach can be taken to avoid excessive costs due to over- or under-design.

Machine Performance Data Feedback Plan

During the concept and development phases the manufacturing engineering and maintenance people, in cooperation with the supplier, should specify the data collecting systems (hardware/software) to be designed into the manufacturing machinery and equipment. The system should have the ability to interface with the plant's overall R&M feedback system and should provide relevant data to the supplier. Appendix D contains further information regarding data collection and feedback.

SUPPLIER RESPONSIBILITIES

During the concept phase and prior to contract award, the potential supplier should understand the evolving R&M requirements. In the proposal and during proposal related presentations, the supplier should be prepared to address how it will meet the R&M qualitative and quantitative requirements dictated by a tailored R&M matrix. These preparations should manifest themselves in the form of an R&M implementation plan to be delivered after contract award.

R&M ACTIVITIES CHECKLIST

Table 3-1 provides an example of a checklist for R&M program activity during the concept phase (Phase 1). This checklist can be employed by both the user and the supplier.

Table 3-1. R&M Activities Checklist for the Conceptual Phase (Phase 1)

Checklist		Supplier	User
S	Safety	X	X
RM	Continuous Improvement Monitor		Me
RM	Equipment Failure Mode & Effect Analysis (FMEA)	X	X, Me, Ma
RM	Process FMEA		
RM	Design Review	X	X, Op, Me, Ma
RM	Machine Data Feedback Plan	X	X, Me, Ma, Op
RM	MTBF	X_E	X_E, Me, Ma
R	Fault Tolerance	X	X
R	Fault Tree Analysis		
R	Life in Terms of Throughput	X	X, Me
RM	Overall Equipment Effectiveness	X_E	X_E, Me, Ma, Op, Pu
R	Reliability Block Diagram	X	X, Me
R	Failure Mode Analysis		
RM	Quality Function Deployment	X	X
RM	Environment	X	X
RM	Life Cycle Costing	X_E	X_E, Me, Fi
RM	Mean Time-to-Repair	X_E	X_E, Me, Ma
RM	Validation Process		
M	Accessibility		
M	Built-in-Diagnostics	X	X
M	Captive Hardware		
M	Color Coding		
M	User Training		
M	Maintenance Procedures		
M	Modularity		
M	Spare Parts Inventory		
M	Standardization		
SM	Ergonomics	X	X
RM	Computer/Process Simulation	X	X

Legend for Table 3-1
  Function In User's Organization
  Manufacturing Engineer = Me
  Maintenance = Ma
  Purchasing = Pu
  Machine Operator = Op
  Plant Management = Pm
  Financial = Fi

R - Indicates the checklist item affects reliability
M - Indicates the checklist item affects maintenance
S - Indicates the checklist item affects safety
X - Indicates the checklist item is recommended for use during the
      program phase
X_E - Indicates the checklist item can be estimated during the program phase
X_M - Indicates the checklist item can be measured during the program phase

NOTE: Refer to Appendix A for further discussion regarding R&M techniques.

SECTION 4

USER AND SUPPLIER R&M ACTIVITIES IN
THE DESIGN AND DEVELOPMENT PHASE (PHASE 2)

In the design and development phase the user should:

1. **Verify the supplier's capability to undertake the actions specified in the tailored R&M matrix and**

2. **Monitor the supplier's progress through periodic meetings and scheduled design reviews.**

INTRODUCTION

The design and development phase (Phase 2) focuses on equipment design and verification of the capability of the evolving design to meet the R&M requirements specified in Phase 1.

USER RESPONSIBILITIES

During this phase the user should work closely with the supplier to ensure thorough understanding of the specified requirements. The user should review the R&M plan to verify that the supplier has the capability to undertake the actions specified in the tailored R&M matrix. Examples of these tailored matrices are included in Section 7.

During this phase the user should monitor the progress of the supplier in R&M activities informally through periodic meetings and formally through scheduled design reviews as discussed later in this section.

SUPPLIER RESPONSIBILITIES

Design Margins

All machinery suppliers should use some degree of conservatism in the design practices employed by the various engineering disciplines (e.g., electrical, mechanical, fluidics, etc.). Reliability is a function of stress. The greater the stress (e.g., electrical, mechanical, thermal, etc.), the less the achieved reliability. While stress analyses after design can provide evidence of overstress conditions, the recommended proactive approach involves estimating the worst case conditions of an evolving design and providing an appropriate design margin. This design margin in electrical engineering is referred to as derating, in mechanical engineering as a safety factor.

For the industrial environment, derating margins and safety factors should be employed at the component level after taking into account the predominant stresses, including thermal. Machinery designers should employ their own set of design margin criteria to ensure that the strength of a component exceeds the applied stress. The use of adequate derating values is the responsibility of the supplier.

Derating and safety factors must be assessed under worst-case conditions for stress and strength. Strength data are available for many electrical and mechanical components and materials. Stress is unique to each application and requires measurements on similar applications and calculations to account for differences of application.

Maintainability Design Concepts

The R&M plan should include maintainability design features that will minimize both corrective action and preventive maintenance downtime, and reduce the frequency of preventive maintenance actions. Design techniques such as accessibility, modularity, standardization, repairability, testability (including built-in test equipment) and interchangeability should be considered by the machinery designer during equipment design. Also, by understanding how the machinery user will maintain the equipment, the designer can evolve a maintainability design that is in concert with the machinery user's logistics support system. Characteristics that should be considered in designing for maintainability include

- Design for repair with standard tools and test equipment;

- Maintenance manuals that are consistent with the skill levels of those performing the maintenance;

- Preventive maintenance requirements that are compatible with existing schedules and inventories;

- Estimates of time to carry out preventive maintenance actions defined by the supplier;

- Easily identifiable locations for predictive maintenance analysis (e.g., vibration sensors);

- Spare parts lists based on equipment reliability characteristics; and

- Minimum spare parts inventory requirements.

Reliability Analysis and Predictions

Machinery suppliers should perform preliminary reliability predictions to support bid responses and verify that the specified reliability requirement and/or goal is attainable. Reliability predictions can be accomplished using data collected on similar or identical equipment.

For reliability data bases of similar equipment, adjustments to the MTBF should be made to reflect differences in use, complexity and other operating characteristics. Where existing data bases are not available, reliability predictions can be performed by generating a Reliability Block Diagram at the component level. Using information contained in reliability data bases, component failure rate estimates can be incorporated into the reliability block diagram and a machine reliability calculated. As additional industrial component reliability data are acquired, the reliability prediction process will become more refined.

Accelerated Life Test

The accelerated life test may be conducted on critical components as dictated by the user. Test conditions must be accelerated in an appropriate manner so that the failure rate can be translated to normal use condition.

Failure Mode and Effects Analysis/Fault Tree Analysis

Machinery suppliers should perform equipment Failure Mode and Effects Analyses (FMEAs) on their machines. These FMEAs may be required by contract/purchase order. Similar to product and process FMEAs, equipment FMEAs focus attention at the subsystem level for purposes of identifying potential failure modes and their impact on satisfactory performance of the machine. FMEAs should be performed early in the design process to assure that critical failure modes are eliminated from the design and that maintainability procedures are defined for the remaining non-critical failure modes. The FMEA should become the primary tool for developing troubleshooting procedures.

The user should perform process FMEAs on new manufacturing systems and involve the manufacturing machinery and equipment supplier. Typically this analysis would be performed early in the development of the new system with selected plant operators and maintenance personnel, and should be coordinated by the manufacturing engineer. This activity focuses engineering and management attention on high risk areas of the new system and requires upper management support during simultaneous engineering to assure appropriate plant participation.

Fault Tree Analyses (FTA) should be conducted on an as-needed basis. An FTA utilizes an effect-and-cause diagram that uses standard symbols to help define layers and relationships of causes of unreliability. In a manner similar to Fishbone Diagrams, FTAs focus on failure causes from the top down while FMEAs focus from the bottom up. Refer to Appendix A for further information on FMEA and FTA.

Design Reviews

Design reviews should be an integral part of the design and development process. A design review is a formalized, documented and systematic management process through which both the machinery supplier and the user review all technical aspects of the evolving design, including R&M. This process typically involves the review of drawings, sketches, engineers' notebooks, analysis results, test documentation, mockups, assemblies, and other hardware and software depictions of the evolving design.

In order to be effective, the design review must be multi-phased and keyed to the various phases of the design process. For large systems, the machinery supplier as well as the key component suppliers should participate in and be an integral part of the design review process. From a R&M standpoint, the review should focus on the R&M activities consistent with the design evolution. For example, during a preliminary design review, focus should be on selecting design assurance activities while the critical design review should concentrate on the results of the R&M analyses. Table 4-1 lists the review objectives at various program phases.

Not all design review phases are applicable to each machine. The machinery supplier (where required by contract or purchase order) and the machinery user should establish which reviews are required for a particular machine and when they should be scheduled.

Table 4-1. Design Review Objectives

Design Phase	Review Objective
1. Concept	<u>Concept Review:</u> Focus on feasibility of proposed design approach
2. Development and Design	<u>Preliminary Design Review:</u> Validate the capability of the evolving design to meet all technical requirements <u>Critical Design Review:</u> Verify that the documented design and related analyses are complete and accurate
3. Build and Install	<u>Build:</u> Address issues resulting from machine build and runoff testing <u>In-Plant Installation:</u> Conduct failure investigation of problem areas for continuous improvement

R&M ACTIVITIES CHECKLIST

Table 4-2 provides an example of a checklist for R&M program activity during the Design and Development phase (Phase 2). This checklist can be employed by both the user and supplier.

Table 4-2. R&M Activities Checklist for the Design and Development Phase (Phase 2)

Checklist		Supplier	User
S	Safety	X	X
RM	Continuous Improvement Monitor		
RM	Equipment Failure Mode & Effect Analysis (FMEA)	X	X, Me, Ma
RM	Process FMEA	X	X, Op, Me, Ma
RM	Design Review	X	X, Op, Me, Ma
RM	Machine Data Feedback Plan	X	X, Me, Ma, Op
RM	MTBF	X_E	X_E, Me, Ma
R	Fault Tolerance	X	X
R	Fault Tree Analysis	X	X
R	Life in Terms of Throughput	X	X, Me
RM	Overall Equipment Effectiveness	X_E	X_E, Me, Ma, Op, Pu
R	Reliability Block Diagram	X	X, Me
R	Failure Mode Analysis		
RM	Quality Function Deployment		
RM	Environment		
RM	Life Cycle Costing	X_E	X_E, Me, Fi
RM	Mean Time-to-Repair	X_E	X_E
RM	Validation Process		Me
M	Accessibility	X	X
M	Built-in-Diagnostics	X	X
M	Captive Hardware	X	X
M	Color Coding	X	X
M	User Training	X	X, Ma, Op
M	Maintenance Procedures	X	X
M	Modularity	X	X
M	Spare Parts Inventory	X_E	X_E, Ma, Me
M	Standardization	X	X
SM	Ergonomics	X	X
RM	Computer/Process Simulation	X	X

Legend for Table 4-2
Function In User's Organization
Manufacturing Engineer = Me
Maintenance = Ma
Purchasing = Pu
Machine Operator = Op
Plant Management = Pm
Financial = Fi

R - Indicates the checklist item affects reliability
M - Indicates the checklist item affects maintenance
S - Indicates the checklist item affects safety
X - Indicates the checklist item is recommended for use during the program phase
X_E - Indicates the checklist item can be estimated during the program phase
X_M - Indicates the checklist item can be measured during the program phase

NOTE: Refer to Appendix A for further discussion regarding R&M techniques.

SECTION 5

USER AND SUPPLIER R&M ACTIVITIES IN
THE BUILD AND INSTALL PHASE (PHASE 3)

In the build and install phase:

Process variables affecting R&M should be identified and targeted for control during the manufacture, assembly and installation of the equipment.

INTRODUCTION

During the manufacturing and assembly of the machine, care must be taken to avoid degrading the R&M capabilities designed into the equipment. Manufacturing process variables affecting R&M should be identified and targeted for control. Installation related anomalies should be addressed prior to machine start-up to reduce the effect of compounding failures.

USER RESPONSIBILITIES

During Phase 3, the user should actively monitor the machine construction process to identify potential R&M related manufacturing defects. Problems identified during run-off, dry run, or other types of qualification or acceptance testing, should be identified by the user. Supplier-initiated root cause analyses results should then be evaluated. Information collected during supplier or user conducted testing should be used to establish an equipment R&M baseline.

SUPPLIER RESPONSIBILITIES

Machinery Parts

Machinery parts should be selected to optimize overall machine reliability. Characteristics of reliability-sensitive or critical components should be defined by the manufacturing machinery and equipment supplier to ensure flow down to the lower tier suppliers. The R&M plan should identify how suppliers are selected and, where applicable, the qualification process used to verify reliability and integrity of critical components.

Tolerance Studies

Machinery suppliers should routinely conduct studies to ensure that electrical and/or mechanical tolerance stacking does not cause equipment failure or premature wear under the worst-case environmental conditions on the manufacturing floor.

Stress Analyses

For select new designs and existing designs that have been shown to be failure prone, the machinery supplier should conduct stress analyses. Stress analyses use numerical analysis to determine the relationship between the strength of the component and the stress induced by the environment under worst-case conditions. Stress analyses will validate the effectiveness of the design margin criteria employed by the machinery supplier. The areas designated for stress analysis investigation should be documented in the R&M plan.

Dedicated Reliability Testing (Qualification Testing)

Machinery suppliers may be required to perform dedicated testing for purposes of verifying attainment of the R&M requirements stated in the procurement documentation. Test duration under normal operating conditions should be mutually agreed to by the machinery builder and user after assessment of the test objectives and the specified R&M requirements. A typical test program could extend to approximately four times the specified MTBF with all failures documented and maintenance times recorded. All reliability testing should be detailed in the R&M plan.

A preliminary assessment of R&M should be done by the supplier before the machine is ready to begin acceptance testing at the supplier's facility. Important machine parameters should be verified as operating within acceptable limits. This testing should help to uncover problems related to infant mortality.

Reliability Data Collection During Acceptance Testing

As a cost-effective alternative to dedicated testing, reliability data can be collected during acceptance testing at the machinery supplier's facility. This data provides an initial indication of reliability capability. Although test time and failure event data may be limited, information collected can be combined with process validation test results to develop a reliability benchmark. This benchmark then becomes the starting point for reliability growth through continuous improvement.

Reliability Data Collection At The User Plant

R&M data should be collected during acceptance testing at the user's plant to verify that reliability characteristics have not been degraded during shipping and installation. This database can also be combined with the database established during acceptance testing at the supplier's facility.

Root Cause/Failure Analysis

Machinery suppliers should be responsible for ensuring that root cause analyses of equipment failures will be performed by either themselves or their associated component vendors. The results of these analyses will be fed back to the user so that user and supplier can jointly determine how to best resolve any deficiencies. The process by which failure issues are to be resolved should be documented in the R&M plan.

R&M ACTIVITIES CHECKLIST

Table 5-1 provides an example of a checklist for R&M program activity during the Build and Install phase (Phase 3). This checklist can be employed by both the user and supplier.

Table 5-1. R&M Activities Checklist for the Build and Install Phase (Phase 3)

	Checklist	Supplier	User
S	Safety	X	X
RM	Continuous Improvement Monitor		
RM	Equipment Failure Mode & Effect Analysis (FMEA)		
RM	Process FMEA	X	X, Op, Me, Ma
RM	Design Review		
RM	Machine Data Feedback Plan		
RM	MTBF		
R	Fault Tolerance		
R	Fault Tree Analysis	X	X
R	Life in Terms of Throughput		
RM	Overall Equipment Effectiveness		
R	Reliability Block Diagram		
R	Failure Mode Analysis	X	X
RM	Quality Function Deployment		
RM	Environment		
RM	Life Cycle Costing		
RM	Mean Time-to-Repair		
RM	Validation Process	X	X, Me, Ma, Op
M	Accessibility		
M	Built-in-Diagnostics		
M	Captive Hardware	X	X
M	Color Coding	X	X
M	User Training	X	X, Ma, Op
M	Maintenance Procedures	X	X
M	Modularity		
M	Spare Parts Inventory		
M	Standardization		
SM	Ergonomics		
RM	Computer/Process Simulation		

Legend for Table 5-1

Function in User's Organization
Manufacturing Engineer = Me
Maintenance = Ma
Purchasing = Pu
Machine Operator = Op
Plant Management = Pm
Financial = Fi

R - Indicates the checklist item affects reliability
M - Indicates the checklist item affects maintenance
S - Indicates the checklist item affects safety
X - Indicates the checklist item is recommended for use during the program phase
X_E - Indicates the checklist item can be estimated during the program phase
X_M - Indicates the checklist item can be measured during the program phase

NOTE: Refer to Appendix A for further discussion regarding R&M techniques.

SECTION 6

**USER AND SUPPLIER R&M ACTIVITIES IN
THE OPERATION AND SUPPORT PHASE (PHASE 4)**

In the operation and support phase:

The user is expected to implement a system of R&M data collection, analysis and feedback. The supplier uses feedback to improve the R&M of existing equipment designs.

INTRODUCTION

In this phase of the life cycle, equipment has been delivered, installed, and tested, and has begun producing product. Data collection and feedback are very important at this phase. Data collection mechanisms should be in place and agreed upon by all parties. Information collected during this phase should lead to R&M growth and continuous improvement. During this phase preventive maintenance should be performed as agreed. For an R&M initiative to be successful, the manufacturing machinery and equipment and component suppliers must have access to maintenance records and R&M data bases.

USER RESPONSIBILITIES

In this phase the user is expected to implement a system of R&M data collection, analysis and feedback. The efficient collection and feedback of equipment operation data is critical to a successful R&M program. R&M data feedback to the supplier is important for improving the R&M of existing equipment designs. Feedback of data provides a basis for improved R&M in new designs for future programs.

These user responsibilities are best achieved by working in partnership with the supplier. For instance, the sharing and analysis of plant maintenance information is much more effective when performed in a non-adversarial user-supplier relationship. In short, the user's R&M practices should be closely integrated with the supplier's R&M program.

Data needed to calculate MTBF and MTTR characteristics are generated through the operating and maintenance activities. These data must be shared with the supplier to become a major part of any user continuous improvement program. These are among the most important measurements made throughout the machine life cycle, and form the scorecard suppliers will be using for all future manufacturing machinery and equipment.

User support of the supplier's R&M program is further enhanced when the user has a sound maintenance program involving

• Planned maintenance;

• Machine operator participation; and

• A proactive manufacturing engineering discipline well-trained in R&M practices.

Users who do not understand R&M practice or who are not organized to provide important maintenance (MTTR) and failure (MTBF) information will not be able to support a successful R&M program.

The practices proposed by the user will support continuous improvement to current plant operations as well as future manufacturing machinery and equipment designs. The R&M activity is a key factor driving the continuous improvement process. The same practice and data base that supports the R&M activity of the supplier is required for problem identification and resolution for in-plant continuous improvement.

Some of the obvious benefits of an R&M data collection system include

• Increased uptime;
• Detection of failure trends;
• Improved information to support Root Cause Analyses;
• Reliability Growth; and
• Maintainability improvement.

Once a problem area has been identified, failure data is required for root cause analysis. These data are critical to predicting failure rates, trending, performing root cause analysis of repeat failures, and providing focus in the continuous improvement activities. For this reason the maintenance department and the machine operator should keep accurate and timely records of each downtime incident. For each failure, the symptoms and the corrective action taken should also be recorded. The operator can record this information either on a paper machine log or directly into the plant monitoring system via a fault screen. The maintenance data base could be in electronic form, ideally integrated with the production monitoring data base. This data base should be capable of providing supplier-specific data.

The accurate and consistent recording of failures, symptoms and corrective actions is far more important than the medium for data collection (paper or electronic). There are many instances where a paper system is providing accurate data.

The analysis of the data is an essential but time-consuming step. Summary reports need to be generated to identify frequency of faults within a bottleneck area.

The importance of R&M should be reinforced at the user's plant with strong leadership from the plant management team. Integration of R&M practices with plant continuous improvement and maintenance programs requires management attention to assure that

- Training requirements are met;

- Objectives are clearly communicated;

- R&M practices are implemented; and

- User R&M practices supporting supplier R&M activities receive sufficient priority.

SUPPLIER RESPONSIBILITIES

Reliability Growth/Maintainability Improvement

When required by contract/purchase order, machinery suppliers are expected to participate in the reliability growth (continuous improvement) process during user manufacturing operations by assisting in the root cause analysis of all failures that occur. The reliability growth (continuous improvement) process is the most effective technique for reducing warranty costs and improving future products. It is strongly recommended that machinery suppliers participate in the reliability growth process even when it is not contractually required. By identifying failure causes and implementing corrective actions, the inherent reliability of equipment can be improved. The level of participation required will be a function of how quickly the machinery reaches a level of mature reliability.

Failure Reporting, Analysis and Corrective Action System

A significant amount of work has been done in the electronics industry to develop feedback systems. These systems provide for the orderly recording and transmission of failure data from the supplier's plant and user sites into a unitary data base. This data base allows identification of pattern failures and rapid resolution of problems through rigorous failure analysis. This resolution of problems promotes the reliability growth of equipment in the field as well as higher reliability for new equipment. Incorporating maintainability data into the data base allows monitoring of repair performance and the continuous improvement of equipment.

The machinery supplier should develop a means for summarizing and analyzing failure data obtained at both supplier and user plants as a means of promoting R&M improvement activities. Reliability growth is accomplished by closing the loop and using failure data to improve machine designs.

Data Exchange

The machinery supplier and user should jointly develop a system for obtaining user equipment performance data. This data should include MTBF, MTTR, failure event data, reliability growth data, and yield. The data format should be mutually agreed upon and provide sufficient detail so that the machinery supplier can isolate a problem and develop corrective action.

This exchange of data benefits both machinery suppliers and users. The benefits to the machinery suppliers include a clear understanding of their machines' performance at user facilities, insight into the reduction of warranty costs, and improved user satisfaction. The benefits to the machinery user are improved machine performance and a more responsive supplier.

SUGGESTED UNIVERSAL DATA FEEDBACK MODEL

Figure 6-1 shows a minimal flow chart that plants should have established for failure documentation. This documentation allows engineering personnel to communicate with suppliers and explains the suppliers' role in the data feedback system.

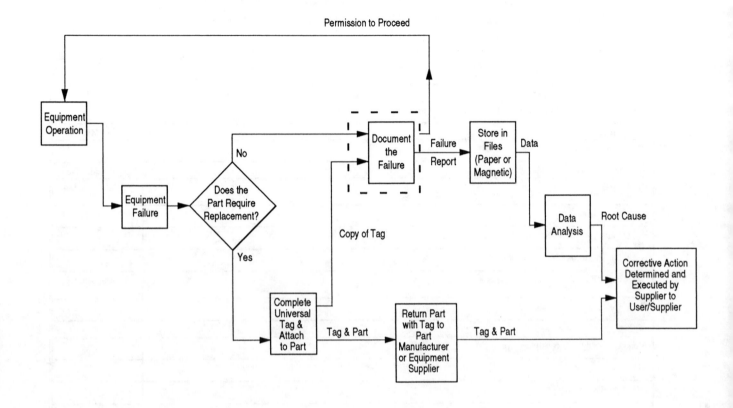

Figure 6-1. Basic Failure Documentation System

R&M ACTIVITIES CHECKLIST

Table 6-2 provides an example of a checklist for R&M program activities during the Operation and Support Phase (Phase 4). This checklist provides examples of items to be considered by both the user and supplier during this phase.

Table 6-2. R&M Activities Checklist for the Operation and Support Phase (Phase 4)

	Checklist	Supplier	User
S	Safety	X	X
RM	Continuous Improvement Monitor	X	X, Op, Ma
RM	Equipment Failure Mode & Effect Analysis (FMEA)		
RM	Process FMEA	X	X, Op, Me, Ma
RM	Design Review		
RM	Machine Data Feedback Plan	X	X, Me, Ma, Op
RM	MTBF		X_M, Me, Op, Ma
R	Fault Tolerance		
R	Fault Tree Analysis		
R	Life in Terms of Throughput		
RM	Overall Equipment Effectiveness	X_M	X_M, Me, Ma, Op
R	Reliability Block Diagram		
R	Failure Mode Analysis	X	X
RM	Quality Function Deployment		
RM	Environment		
RM	Life Cycle Costing		
RM	Mean Time-to-Repair	X_M	X_M, Me, Ma, Op
RM	Validation Process		
M	Accessibility		
M	Built-in-Diagnostics		
M	Captive Hardware		
M	Color Coding		
M	User Training	X	X, Ma, Op
M	Maintenance Procedures		
M	Modularity		
M	Spare Parts Inventory	X_M	X_M, Ma, Me
M	Standardization		
SM	Ergonomics		
RM	Computer/Process Simulation		

Legend for Table 6-2

Function in User's Organization
Manufacturing Engineer = Me
Maintenance = Ma
Purchasing = Pu
Machine Operator = Op
Plant Management = Pm
Financial = Fi

R - Indicates the checklist item affects reliability
M - Indicates the checklist item affects maintenance
S - Indicates the checklist item affects safety
X - Indicates the checklist item is recommended for use during the program phase
X_E - Indicates the checklist item can be estimated during the program phase
X_M - Indicates the checklist item can be measured during the program phase

NOTE: Refer to Appendix A for further discussion regarding R&M techniques.

SECTION 7

R&M AND CONTRACTING

For improved R&M in purchased equipment, the key lies in the planning, specification, and procurement practices.

R&M planning identifies what has to be done, who should do it, and when it should be done.

An R&M matrix is required to define quantitative requirements and spell out the amount of detail needed for each task. The worksheet provides the supplier with a clear picture of the R&M requirements over the equipment life cycle.

The matrix and worksheet should be incorporated into specifications, bid packages, or other procurement documents.

INTRODUCTION

The considerations discussed in this section apply to the user as well as to machinery suppliers. A primary key to improved R&M and reduced cost of machinery operation lies in the planning, specification, and procurement practices for machinery and equipment.

The following sequence of actions should be pursued by the user and supplier during the planning, specification, and procurement processes.

1. Reach management-level agreement on R&M requirements, fully recognizing that decisions will impact all aspects of machinery acquisition and operation.

2. Distribute written R&M requirements, and plans to reach those requirements, to responsible personnel (e.g. engineering , purchasing, plant supervisors, etc.) at the user plant and to suppliers.

3. Bring top management together with all responsible personnel to discuss R&M requirements to assure that there is understanding and acceptance of the goals.

4. With the overall plant project requirements established, allocate R&M requirements to all machinery subsystems.

5.　　Create R&M measurement criteria (e.g. test procedures, data requirement, and verification).

6.　　Include potential suppliers in discussions of modified procurement practices for manufacturing machinery and equipment to achieve R&M requirements.

7.　　Incorporate R&M documents and criteria into a Supplier Bid Package and send out Requests for Proposals to selected suppliers. Tailor the R&M plan as appropriate for the equipment under consideration.

8.　　Respond to further supplier questions.

9.　　Analyze potential suppliers and rate them relative to their R&M practices.

10.　　Receive bids and place orders with suppliers who meet the bid criteria.

PLANNING

R&M planning determines how to achieve the R&M goals necessary to meet operational requirements for machinery. The plans identify what has to be done, who should do it, and when it should be done. There must be a process for designing for R&M at the start, coupled with a process for ensuring that inherent R&M is achieved during the design/development, build/install, and operation phases of manufacturing machinery and equipment.

The thoughtful and thorough development and communication of specifications is the most fundamental element of R&M assurance. All subsequent design decisions, manufacturing activities, and service systems are based on specifications, in whatever form they take. A specification, in this context, refers to a description of what the machinery is supposed to be and do, not how to design, build or test it. It is important to realize that if machinery is not designed for R&M, it may be prohibitive to build R&M in later.

The Simultaneous Engineering approach allows incorporation of new disciplines such as design assurance techniques and engineering analyses in order to determine true LCC (see Appendix C). This, in fact, involves the difficult task of wisely investing time and effort as opposed to simply expending large amounts of capital. Under this structure, a cross functional team is formed consisting of a product design engineer, process engineer, and representatives from purchasing, manufacturing, quality control, financial representatives, and the selected supplier's engineering representatives.

During the design iteration process, the team has repeated opportunities to enhance the R&M of the equipment and make cost-benefit trade-offs.

The team continuously reviews and refines the product and process design with the absolute requirement that no open issues regarding product, manufacturing or reliability exist at the conclusion of the planning period. The net result is a design and process that all members of the team fully understand and support.

R&M Matrix

The procurement package issued by the user should include two items related especially to R&M: the tailored R&M matrix and the R&M worksheet. The tailored R&M matrix should be developed by the technical activity procuring the equipment. The purpose of tailoring is to define an application-specific program for the equipment being procured as a function of both supplier R&M capability and the amount of design engineering required to meet specified performance requirements. As an example, Figure 7-1 provides an R&M matrix tailored for the procurement of equipment (stand-alone machines) with only minor modifications. In addition to defining R&M quantitative related requirements, specific tasks are required during each of the remaining life cycle phases. As can be seen, reliability predictions and equipment-level FMEAs are required during the Design and Development phase. The detail required in each of these tasks must be specified in the R&M worksheet. For example, if reliability predictions are only required for certain subsystems on the machine (e.g. electrical control, hydraulic, etc) then that level of detail must be stated in the R&M worksheet that accompanies the tailored matrix.

Figure 7-2 is another example of tailoring R&M requirements for a specific procurement. This acquisition is for an existing supplier designed machine requiring major modifications to meet user needs. As can be seen, the number of R&M tasks has increased over the previous example. Once again, an R&M worksheet is required to define quantitative requirements and spell out the amount of detail required for each task specified. In this example, the user has dictated that certain tasks are required only for the modified design. The user has also dictated that only one design review is required.

The third example is shown in Figure 7-3. This matrix, along with a R&M worksheet, provide the supplier with a clear picture of the R&M requirements over the equipment life cycle.

R&M Program Planning Worksheet

A R&M Program Planning worksheet (see Figure 7-4) should accompany each tailored R&M Program matrix. The purpose of the worksheet is to provide specific details relating to the program elements specified in the R&M matrix. For example, Figure 7-3 indicates Reliability Qualification Testing is required in Phase 3, Build and Install. The R&M Program Planning Worksheet should then detail on what components and subassemblies the Reliability Qualification Testing should be performed.

The following three examples provide clarification of how R&M quantitative and qualitative requirements are included in the specification. It will be up to the user, with help from the supplier community, to ensure that their documents are properly tailored to be consistent with user product requirements to avoid placing too little or too much emphasis on R&M.

R&M Program Matrix For Suppliers

Existing Design or
Minor Modifications
Single Machine Operation

	Concept – Phase 1	Design & Develop – Phase 2	Build & Install – Phase 3	O & S – Phase 4	Disposal – Phase 5
Reliability Requirements	X				
Maintainability Requirements	X				
Failure Definition	X				
Environment/Usage	X				
Design Margin					
Maintainability Design					
* Reliability Predictions		X			
* FMEA/FTA		X			
Design Reviews					
Parts					
Tolerance Studies					
* Stress Analysis			X		
R Qualification Testing					
R Acceptance Testing					
R Growth/M Improvement				X	
Failure Reporting				X	
Data Feedback				X	

* If not previously performed, corrective action should be taken for high failure rate items and must be taken for safety-related items.

Figure 7-1

R&M Program Matrix For Suppliers

Existing Design with Major Modifications

	Concept – Phase 1	Design & Develop – Phase 2	Build & Install – Phase 3	O & S – Phase 4	Disposal – Phase 5
Reliability Requirements	X				
Maintainability Requirements	X				
Failure Definition	X				
Environment/Usage	X				
* Design Margin		X			
* Maintainability Design		X			
** Reliability Predictions		X			
** FMEA/FTA		X			
* Design Reviews (one critical design review)		X			
Parts					
Tolerance Studies					
Stress Analysis			X		
R Qualification Testing					
R Acceptance Testing					
R Growth/M Improvement				X	
Failure Reporting				X	
Data Feedback				X	

* Modified design only.
** Modified design and existing design if not previously performed.

Figure 7-2

R&M Program Matrix For Suppliers

New Concept and Design (All Complexities - Single Machine to Transfer Line)

	Concept – Phase 1	Design & Develop – Phase 2	Build & Install – Phase 3	O & S – Phase 4	Disposal – Phase 5
Reliability Requirements	X				
Maintainability Requirements	X				
Failure Definition	X				
Environment/Usage	X				
Design Margin		X			
Maintainability Design		X			
Reliability Predictions		X			
FMEA/FTA		X			
Design Reviews (minimum of 3: concept, preliminary, critical)		X			
Parts			X		
Tolerance Studies			X		
Stress Analysis			X		
* R Qualification Testing			X		
R Acceptance Testing (optional)			X		
R Growth/M Improvement				X	
Failure Reporting				X	
Data Feedback				X	

* Critical components and/or assemblies.

Figure 7-3

R&M PROGRAM PLANNING WORKSHEET

Date: _____

1. Facility/Machine Name: _____

2. Process/Operation Number: _____

3. Brass Tag Number: _____

4. Supplier: _____

5. Model Year: _____
   Plant: _____
   Vehicle: _____

6. Identify R&M concerns on similar facility/machines now in operation.

   _____
   _____
   _____
   _____
   _____

7. List the pertinent environmental factors affecting R&M and machine performance:

Environmental Factor	Range	Rate of Change

8. Identify R&M concerns for this application, from similar applications, and from environmental considerations:

   _____
   _____
   _____
   _____

Figure 7-4

9. List the R&M target values for this machine/facility:

Factor	Requirement
Expected Life	
MTBF	
MTTR	

10. For each equipment life cycle phase, list the R&M tools and technqiues to be used on this program:

PHASE 1 - CONCEPT

Tool/Technique	Responsible Personnel

PHASE 2 - DESIGN & DEVELOPMENT

Tool/Technique	Responsible Personnel

PHASE 3 - BUILD & INSTALL

Tool/Technique	Responsible Personnel

PHASE 4 - OPERATION & SUPPORT

Tool/Technique	Responsible Personnel

PROCUREMENT

A general checklist of purchasing practices is as follows:

• Develop a cooperative relationship with suppliers that establishes performance levels before prices are negotiated.

• Get suppliers of key items involved with designers early in the machinery development process.

• Establish a formal supplier rating system.

• Identify the relative importance of critical characteristics to suppliers.

• Include R&M requirements in purchase specifications with a description of how the requirements are to be verified.

• Share performance data with suppliers as an aid to technical problem solving.

• Qualify critical processes used by suppliers using statistical principles prior to start-up.

• Conduct failure and root cause analyses by multifunctional joint user/supplier teams.

In response to the user's request for quotes, it is expected that machinery suppliers will define an R&M plan that will be implemented in their facility during the design/development and build/install phases of the manufacturing machinery and equipment life cycle. The plan is expected to show how the supplier will integrate the various R&M techniques, disciplines, and procedures that will result in the supplier meeting the quantifiable R&M requirements.

Full compliance in implementing R&M practices may not be easy and it may take time to achieve. However, the degree, extent and reasonability of supplier responses will be heavily weighed in order-placement decisions.

Supplier R&M implementation plans should generally adhere to this R&M guideline.

APPENDIX

APPENDIX A

R&M TOOLS AND TECHNIQUES

This appendix provides further information about some of the tools and techniques referenced within the body of the guideline. This appendix does not provide the level of detail that a potential reliability engineer will need.

MAINTAINABILITY

Safety

Safety engineering must be introduced at the design stage, not after the equipment is built. Safety personnel must be consulted up-front to utilize fully the best technology available in a safe and ergonomically efficient manner. Properly designed, the operator's environment will not only reduce the risk of injury, but will also avoid exposure to health risks or activities likely to cause repetitive motion disorders. Pinch points guarding, safety labels, personnel guards, warning devices, lock-outs and other appropriate safety measures must be integrated into the design. Safety requirements must be included in the specifications.

All applicable safety standards must be adhered to.

Accessibility

Accessibility means having sufficient working space around a component to diagnose, troubleshoot, and complete maintenance activities safely and effectively. Provision must be made for movement of necessary tools and equipment with consideration for human ergonomic limitations.

Diagnostics

Diagnostic devices indicating the status of equipment should be built into manufacturing machinery to aid maintainability support processes.

The diagnostics can be as simple as a visual display indicating the equipment's status as a go/no-go condition, or as sophisticated as a knowledge-based expert system with the capability of analyzing a problem and recommending the most likely solution.

Diagnostic systems should have the capability of storing equipment performance data as permanent records for reliability analysis and supplier feedback. This system should support the reliability growth management process. Output from diagnostic systems should be in a format that is compatible with commercially available data base management software.

When component assemblies and subsystems are used to create a manufacturing system, hardware and software "hooks" should be put in place in the concept and design phase to facilitate integration of the diagnostics system in the build phase.

Diagnostic systems should indicate the specific component to replace or repair.

Captive Hardware and Quick Attach/Detach

Captive and quick attach/detach hardware provides for rapid and easy replacement of components, panels, brackets and chassis. The environment in which these devices are used may restrict the type of device used. Spare parts and replaceable subassemblies should also be configured with these devices preassembled. Examples include

- plate, anchor and caged nuts;
- push and snap-in fasteners;
- clinch and self-clinching nuts; and
- quarter-turn fasteners.

Visual Management Techniques

Visual Management techniques differ for varying types of equipment. A team effort must exist between supplier and user to deliver the best techniques to the user. These should be reviewed on a continual basis at concept/design, during machine build, and on the manufacturing floor by all the team members. Visual management techniques are used on machinery and equipment to bring the workplace awareness to a level that allows problems and abnormal conditions to be recognized quickly at a single glance. A visual management system enhances the equipment inspection process by allowing quick identification of safety, quality, environmental, equipment, and process abnormalities.

Typical visual management techniques include

- Match marking of all fasteners (nuts, bolts, screws, etc.), whether fixed, adjustable, or critical;
- Match marking of all control adjustments (pressure, flow, temperature, speed, level, voltage, current, etc.);
- The identification of normal operating ranges and levels;
- Direction of flow and product color coding on piping and hoses;
- Direction of rotation (drives, belts, chains, motors, etc.);
- Function labels (switches, valves, buttons, lights, etc.);
- Identification labels (cabinets, panels, boxes, etc.);
- Filters (lube, hydraulic and air) that indicate when dirty;
- Filters labeled with replacement filter element number;
- Belt and chain drives with guarding that permits quick visual inspection and

access;

- Replacement belt or chain number labels on guarding;
- Each lube point labeled with product number and color code;
- Temperature sensitive labels on all critical components (motors, drives, controls, hydraulic units, etc.);
- Equipment layout with all electrical control panel safety lockout points indicated (affixed to the main electrical control panel);
- Equipment layout with all lubrication fill points, frequencies and product codes indicated (affixed to the main electrical control panel);
- The identification of all control drawing numbers on the main electrical control panel;
- Signals or alarms that indicate a major abnormality, safety interlock tripped, process out of control, etc.; and
- Equipment and process operator inspection list (affixed to the main electrical control panel).

Spare Parts Management

Maintenance of manufacturing machinery and equipment requires a readily available supply of spare parts and supporting materials to operate, maintain, and service the equipment. Spare parts management will identify and make available the required quantities of spare parts at an optimum inventory cost to the equipment user.

Plans for equipment support through spare parts management should begin during the equipment design phase and continue through the life cycle of the equipment. Consideration should be given to the lead time required to requisition, manufacture and receive into inventory the required parts or materials to avoid the excessive costs of procuring replacement parts on an emergency basis.

The manufacturing machinery and equipment supplier should make a recommended spare parts list available to the equipment user. Sourcing of spare or replacement parts, including consumable materials, should be managed to ensure that the performance and capability of the manufacturing machinery and equipment is maintained at or above the original manufacturer's specifications.

Maintenance Procedures

Maintenance procedures must describe in detail the adjustments, replacement, and repair of machine systems, subsystems, and component parts. The original equipment manufacturer will provide recommended preventive maintenance procedures at intervals based on time or machine cycle count. Maintenance requirements should be prioritized to enable the equipment user to prioritize maintenance scheduling related to the criticality of the activity.

The maintenance procedures should be contained in service manuals or a computerized data base reflecting the specific content and configuration of the equipment being supported. Exploded view illustrations, photographs, simplified assembly drawings, and parts lists relating to the required maintenance activities and procedures should be included wherever applicable. Technical information such as pressure settings, operational sequences, and moving part clearances should be included as appropriate.

Modularity

Modularity requires that designs be divided into physically and functionally distinct units to facilitate removal and replacement. Modularity mandates design of components as removable and replaceable units for an enhanced design with minimum downtime. Modular design concepts typically are thought of in terms of electrical black boxes, printed circuit boards and other quick attach/detach electrical components. These concepts are also applicable to the mechanical elements of production equipment.

Modularity offers several advantages:

- New designs can be simplified and design time can be shortened by making use of standard, previously developed building blocks.

- Specialized technical skill will be reduced.

- Training of plant maintenance personnel is easier.

- Engineering changes can be made quickly with fewer side effects.

RELIABILITY

Reliability Growth Management

Reliability Growth Management involves the careful identification and cataloging of failures and operation times as a development program progresses. Cumulative data are plotted to observe a trend line. Instantaneous measures of interest are estimated from the trend line parameter, and forecasts are made by extrapolation.

Usually these trends will form a fairly straight line when plotted on log-log paper.

Root Cause Analysis and Failure Analysis

Root Cause Analysis is a logical, systematic approach to identifying the basic reasons (causes, mechanisms, etc.) for a problem, failure, non-conformance, process error, etc. Whenever a significant problem occurs (i.e. low frequency of occurrence with high cost in time and/or money, or high frequency of occurrence), Root Cause Analysis should be implemented. The result of Root Cause Analysis should always be the identification of the basic mechanism by which the problem occurs and a recommendation for corrective action. "No corrective action required" is an acceptable recommendation for corrective action when properly justified. Root Cause Analysis cannot be closed out until all required corrective action has been developed and implemented. Many users, as well as professional societies (SAE, ASQC, IEEE, etc.) and the U. S. Government, have developed methodologies for performing Root Cause Analysis. When Root Cause Analysis is performed, all of the investigation, analysis, and results (including corrective action) must be completely documented. This documentation should be entered into the R&M Data Collection System where it can be used to help prevent, diagnose, or mitigate future problems.

Failure Analysis is a special case in Root Cause Analysis where a physical failure has occurred. Failure Analysis is defined in the glossary and is synonymous with terms such as "Physics of Failure", "Reliability Physics", etc. Failure Analysis usually involves laboratory analysis and produces data that can include measurements of the failed item and analytical photographs using equipment such as X-ray machines, scanning electron microscopes, spectrographs, and optical microscopes. Proper preparation and preservation of samples and documentation of results can allow for independent expert evaluation of the analysis when disputes arise in determining the failure mechanism or the root cause of the failure mechanism.

Examples of disputes include

- Defective part or part misapplication, and

- Overstressed part application or equipment operated outside of design range.

Like Root Cause Analysis, Failure Analysis is not complete until all recommended corrective actions have been developed and implemented.

Failure Mode and Effects Analysis

A Failure Mode and Effects Analysis is one of the more frequently used techniques in reliability engineering. A complete FMEA program should span the entire life cycle of the manufacturing equipment, using feedback from the data collection process to update the FMEA materials for reliability growth management support and reference for future engineering design analysis.

A Failure Mode and Effects Analysis (FMEA) involves evaluating a design or process through the following steps, and creating rankings to prioritize the related issues.

1. Identify potential failure modes and effects of the failure, and the expected seriousness of failure.

2. Determine the potential cause and probability of occurrence of the anticipated failure.

3. Identify existing controls that are designed to prevent the failure from occurring without detection or to detect the failure and initiate corrective action. Determine the likelihood of detecting the onset of failure before it occurs.

4. Identify actions required to prevent, mitigate or control failures, improve the likelihood of detecting the failure if it does occur, or eliminate the failure.

5. Assign an individual or group the responsibility for performing the recommended action and documenting the action taken to implement the recommended action.

The purpose of FMEA is to analyze a product or process design to ensure that it meets design and operational requirements. As potential failure modes are identified, corrective action can be initiated to eliminate them, continuously reduce their potential for occurrence, and to detect the failure as soon as possible to minimize or eliminate subsequent damage. FMEA also documents the rationale used for the design or process.

FMEA summarizes the thoughts of a cross-functional group of "experts" in evaluating items that could conceivably go wrong based on experience and past problems. This systematic approach parallels and formalizes the mental discipline that an engineer normally goes through both to develop application requirements and to solve problems that occur.

FMEA utilizes a combination of occurrence, detection probability, and severity criteria to develop a Risk Priority Number (RPN) to prioritize corrective actions. A disciplined review and analysis of a new or revised machine component is used to anticipate, resolve and/or monitor potential problems during the planning stages of a machine component design or machining process.

The process FMEA should be performed during the development of the manufacturing process, either during the equipment proposal definition phase or during the simultaneous engineering or pre-engineering phase, whichever is relevant. Design or equipment FMEAs should be integrated into the design development process to enable the results of the analysis to be considered prior to manufacturing the product.

Commercially available FMEA documentation software, formats and forms should be utilized to simplify training and to commonize the procedures and document management.

Reliability Block Diagrams

Reliability block diagrams are planning tools used to recognize how each equipment component contributes to the functional reliability of the equipment. The reliability block diagram used during the initial phases of development/design lays out the equipment design based on component interactions.

Reliability block diagrams are used to calculate system reliability based on component reliabilities. They can also be used to determine the minimum component reliability necessary to achieve a desired system reliability.

Series Systems

In a series system, all components must operate successfully for the system to function correctly. The reliability block diagram for a series system consisting of (n) components is given by:

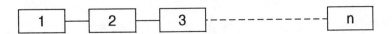

The reliability equation for calculating system reliability (R_s), in a series system is:

$$R_s = R_1 \times R_2 \times R_3 \times ... \times R_n$$

where R_i is the reliability of the ith subsystem.

Example: Automotive Engine Component Transfer Line. An automotive engine component transfer line has 12 station elements plus 3 serially-related support elements (15 total serial interactions).

If each element has a reliability of 0.98, what will be the transfer line reliability?

R_s = 0.98 x 0.98 x 0.98 x 0.98x 0.98 x 0.98 x 0.98 x 0.98 x 0.98 x 0.98 x 0.98 x 0.98x 0.98 x 0.98 x 0.98 = 0.7386

Parallel Systems

A system is considered to have parallel elements if any of the parallel elements can fail without degrading the performance of the system. A system may be considered partially parallel if the failure of a parallel component degrades the system performance to a predictable level, but does not prevent the system from functioning. A parallel system is not considered to have failed unless all redundant subsystems have failed.

The reliability equation for calculating system reliability (R_s) in a parallel system is:

$$R_s = 1-[(1-R_1)x(1-R_2)x\ ...x(1-R_n)]$$

where R_i is the reliability of the ith subsystem.

Example: Belts Driving a Drilling Spindle. The failure of any one of three belts could occur without causing a failure of the drilling process while the failure of any other single component will cause a process failure. The belt-driven drilling spindle has three identical drive belts.

If each belt has a reliability of 0.95, what will be the total belt drive reliability?

$$R_s = 1 - [(1-0.95) \ x \ (1-0.95) \ x \ (1-0.95)] = 0.999875$$

Manufacturing machinery and equipment is frequently a combination of serial and parallel components and processes. The analysis of complex combinations of serial and parallel elements uses the previous formulas in combinations that mirror the equipment configuration to calculate system reliability.

Computer Simulation

Computer simulations are descriptive modeling techniques used to study manufacturing systems to confirm system performance expectations. The simulation process should begin prior to the construction of the manufacturing machinery and equipment in Phases 1 and 2, to enable the findings of the process to be implemented in the system design. The need for simulation varies depending on the complexity and expectations of the system.

Simulation programs should be a joint effort between the equipment supplier and user. Full support should be provided by manufacturing management.

The objective of a simulation analysis is to determine the predicted performance of the total manufacturing system in response to bottlenecks, breakdowns and resource constraints within the system. By optimizing the process flow, selecting and sizing automation buffers, and specifying support resources in accordance with the recommendations from the simulation team, system efficiencies will be improved and costs reduced. With accurate data and careful configuration of the model, the system can be operated "in production" before it is built.

Potential areas for optimization include

- Throughput of components,
- System efficiency,
- Location of queues within the system,
- Location of production bottlenecks,
- Inspection frequencies,
- Tool change management recommendations,
- Batch sizes for multiple part systems,
- Maintenance management program recommendations,
- System manpower requirements, and
- Manpower requirements by trade classification:
  - Operators,
  - Repairmen,
  - Quality inspectors.

Computer simulation support data must be carefully chosen and applied. The simulation should become a supporting process within the manufacturing system and should be updated as often as required to accurately depict the system configuration. Management decisions regarding recommendations from the reliability growth management process can be modeled at a minimal cost prior to implementing equipment changes to determine the net effect of the changes without making a final commitment of resources (manpower and capital).

Fault Tree Analysis

Fault Tree Analysis (FTA) is a top-down method of systems analysis used to explore the possible occurrence of undesirable events. FTA methodology starts with the undesirable event and logically evaluates the causal interrelationships within the total system. These interrelationships determine the conditions that must be present to cause the undesirable event to occur.

A fault tree is one of the few techniques that can portray the interaction of many factors and handle the need to consider several failure conditions that, occurring at once, may trigger the undesirable event. Fault trees include software and human errors, whereas FMEA is generally for hardware and stand alone failures. All safety aspects of the design must be analyzed using FTA early in the design phase to allow changes to be made quickly and easily.

Logic Symbols

Fault trees are constructed using common logic symbols to portray the situation graphically. The *AND* and *OR* gates are the most used symbols and are depicted graphically in the table below. The *AND* gate is used to represent an output condition that can only occur if all of the input events have occurred. The *OR* gate represents a condition where the output condition can occur if any one of the input conditions is present. Other logic symbols are available to represent different logical relationships.

Table A-1: Fault Tree Logic Gates

Gate Symbol	Gate Name	Casual Relationship
	AND	Output event occurs if all the input events occur simultaneously
	OR	Output event occurs if any one of the input events occur

Event Symbols

The event symbols are the next building block needed in constructing a fault tree. A few common event symbols are shown in Table A-2. The top event is at the top of the tree and is the focus of the FTA. This event contains a description of the system-level fault.

A rectangle indicates output events that occur as the fault tree is developed.[1] These events will have logic gates as inputs. The fault tree will end with events represented as circles or diamonds. The circles represent basic faults that cannot be further reduced. The basic fault represented by the circle will be a clearly defined component failure and will need no further expansion. The diamond represents an undeveloped event. This event may not be fully developed because of lack of significance or information.

Table A-2: Fault Tree Events

Event Symbol	Event Name	Description
⬭	Top	Contains the description of the system-level fault or undesirable event
▭	Intermediate fault	Contains a brief description of a lower level fault
△	Input	Contains an input fault to the system; this input can be either a condition from a source outside the system or the normal failure of a system component
○	Basic	The lowest level of event under investigation
◇	Undeveloped	An event which we do not wish to expand further

---

[1]Some analysts also use a rectangle for the top event.

A simple fault tree for a light circuit is shown in Figure A-1. Here any of the system faults listed below the *OR* gate will cause the top condition to occur.

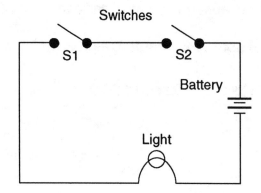

Figure A-1a: Simple Light Circuit

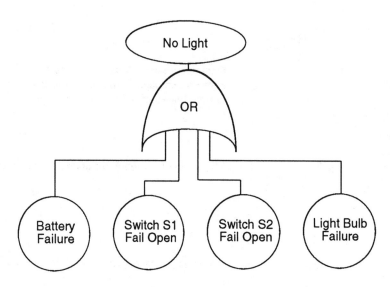

Figure A-1b: Fault Tree for Light Circuit

Figure A-1. Fault Tree Example of a Simple Light Circuit

Environmental Considerations

The R&M process must address the plant environment when considering the design criteria for any process or piece of equipment. The environmental conditions that the process or equipment will be expected to encounter must be thoroughly documented, including not only the levels but also the rates of change. Some environmental factors are

- Temperature,
- Mechanical shock,
- Immersion or splash,
- Electrical noise,
- Electromagnetic fields,
- Ultraviolet radiation,
- Humidity,
- Corrosive materials,
- Pressure or vacuum,
- Contamination and their sources,
- Vibration, and
- Utility services.

In addition to environmental conditions that could affect the process or equipment, considerations must be given to the effects that the process or equipment will have on the plant, the work place, and the earth's environment. The design of any process or piece of equipment must comply with EPA (Environmental Protection Agency) regulations and strive to eliminate totally the use of hazardous materials and the generation of any by-products that can adversely affect the work place. Some of these are

- Coolant/splash, mist, residue;
- Water/splash, spray, residue;
- Oil (lube & hydraulic);
- Welding flash;
- Heat & cold;
- Mechanical shock;
- Electrical noise;
- Electromagnetic fields;
- Cleaning solvents;
- Cutting chips;
- Dust, dirt, sand;
- Mishandled process parts;
- Humidity;
- Vibration;
- Noise levels; and
- Ultraviolet radiation.

Ergonomic Considerations

Ergonomics is an applied science that deals with the interface between people and their work place. Ergonomics takes into consideration the characteristics and capabilities of people in the design and arrangement of equipment, work place, tools, work methods, and facilities, so that people and things will interact most effectively and safely. Some of these considerations are

- Maintainability,
- Operability,
- Ease of inspection (to detect abnormalities), and
- Access points.

APPENDIX B

R&M TRAINING

INTRODUCTION

Training of the OEM and component supplier's engineering and manufacturing personnel and the user's operators and skilled crafts maintenance personnel is essential in retarding deterioration of production machinery.

There are two categories of training required:

- Awareness training in R&M "basics";

- Detailed training in R&M application to specific tasks.

There are also two focuses of the training:

- Train the supplier's design engineers and the user's manufacturing engineers to use the design tools (FMEA, FTA, etc.) needed to assure that R&M is "designed-in" to the manufacturing machinery and equipment;

- Train the user to determine what, how, and when maintenance functions are to be conducted.

Business plans of both supplier and user companies must include the commitment of resources for their training needs. Training should be planned to occur when it can be used to its fullest potential. For suppliers, training should occur when the practices can be implemented on new design projects or be effectively incorporated into current projects. For users, the time that training occurs depends on the activities involved. Training for user management and purchasing personnel should occur prior to ordering any equipment. Training the operators should occur when the equipment is installed on the plant floor. It is useful to consider training as a "Just-In-Time" (JIT) item. Training is most effective when it can be used immediately.

Table B-1. Recommended R&M Training Matrix

Topic	Supplier					User				
	Functional Mgmt.	Sales & Applic. Engr.	Design Engr.	Mach. Builder	Service Engr.	Functional Mgmt.	Manuf. Engr.	OPS & Maint.	User Purchase	User Finance
FMEA FTA	A	D	D	A	A	A	D	A	A	
QFD	A	D	D		D	D	D	D		
Derating	A	D	D	A	A	A	D	A		
Other Rel. Tools	A	D	D	A	A	A	D	A		
LCC	A	D	D			D	D		D	D
Qualification Test	A	D	D	A		A	D	D	A	
Acceptance Testing	A	D	D	D		A	D	D	A	
Rel. Growth Monitoring	A	D	D		D	A	D	D		
Machine Qualification	D	D	D			A	D	D	A	
Data Collection	A	A	A	A	D	A	D	D	A	
Failure Analysis	A		D		D	A	D	D	A	A
Diagnostic Training			D	A	D		A	D		
Periodic Maint. Reqmts.		A	D		D		D	D		
Machine Operation			D		D	A	A	D		

- A - Suggest Awareness Training
- D - Suggest Detailed Training

R&M 'BASICS' TRAINING

Management Commitment

The executive management of both supplier and user companies must be cognizant of and committed to the principles and application of R&M. The management of each company subscribing to this R&M Guideline should utilize it as a training vehicle and a constant reference to their commitment.

Management should integrate R&M into the daily jobs of personnel, starting with the suppliers' machinery designers and extending to the plant operator.

Multifunctional Training

R&M techniques and practices cover many disciplines and functional departments within the participating organizations. Suppliers need to be primarily concerned with providing education that includes R&M design of tools and quality instruction to the user's maintenance personnel. Users need familiarity with design tools to ensure that the tools are correctly applied. Even more important is the training of operations and maintenance personnel in the care of their equipment throughout its life cycle.

Table B-1 suggests a training program for both supplier and user. The points discussed in the following paragraphs expand on the information in Table B-1.

Supplier Responsibility

- Supplier management needs to understand R&M tools and requirements so they can direct the application of R&M and provide guidance and communication to their subordinates.

- Sales and application engineers need to understand R&M requirements in bid specifications, proposals and sales contracts.

- Design engineers are the key to providing inherent R&M of machinery and equipment. What they "design-in" to their product greatly influences life cycle cost to the user. The training of design engineers in R&M must be thorough and continuously improved. They must understand their duty to use R&M tools and provide the technical instruction and data for continued maintenance of their product.

- The machine builder, a skilled craftsman, must be familiar with the concepts of R&M in order to execute the engineer's design.

- Frequently, the service engineer and the service department are the principal interface between the supplier and user after machine installation. They are responsible to communicate knowledge of R&M principles such as MTBF, MTTR, data collection, maintainability issues, and root cause failure analysis to design engineers. The service engineer invariably spends more time in the user plant than any other supplier representative and becomes familiar with what works and what does not work. Total training in R&M issues and principles will improve the flow of information, resulting in continued improvement to machinery.

User Responsibility

- Just as supplier management understands and commits to R&M principles, user management must also commit. It is the responsibility of user management to provide direction and guidance to their subordinates and communicate the logic and justification for R&M requirements to purchasing and finance.

- Manufacturing engineers need to understand reliability tools to judge the degree they must specify. The manufacturing engineers along with the supplier's design engineers are the fulcrum of any R&M effort, and therefore, their training is critical.

- It is the user's operators and maintenance personnel who utilize the opportunity for R&M that the design engineer has "designed-in" to the machine. They require operation- and maintenance-specific training to provide proper use and care of the machine. They also need to understand their role in data collection and failure analysis.

- Purchasing organization representatives need to be aware of R&M requirements to be able to understand their importance in the contract administration process.

- Finally, finance personnel must validate the LCC data and assist in financial trade-off analysis.

APPENDIX C

R&M AND LIFE CYCLE COSTS

LIFE CYCLE PHASES

The life cycle phases of machinery and equipment refer to the period of time from system concept and definition to disposal and are usually identified as follows:

1. System Concept and
   Definition
2. Design and Development   } Non-Recurring Cost
3. Manufacturing, Building
   & Installation

4. Operation/Support   } Support Cost
5. Conversion/ Decommission

Life cycle cost (LCC) is the total cost of ownership of a system during its operational life. A purchased system must be supported for its total life cycle. The cost of support over the life cycle is usually much more than the initial acquisition cost. Acquisition cost is primarily concerned with the first three phases resulting in installation of the equipment in the plant. This initial acquisition cost is a non-recurring cost factor while support cost goes on until system conversion/decommission. Up to 95% of the total LCC is determined by decisions made during the concept and design phase.

The purpose of LCC analysis is to explore various alternatives to identify the most cost-effective production machinery and equipment for a specific application. Applying the LCC concept simply means identifying and summing all costs associated with the system's life cycle.

APPLICATION OF THE LIFE CYCLE COST CONCEPT

Application of the LCC concept requires the following steps:

1. Analyze and define the total life cycle phases for the system.

2. Decide on a satisfactory life period.

3. Determine the cost factors and relationships for each of the phases.

4. Consider the time related factors for costs such as inflation and rate of return.

5. Calculate the LCC by formulating a mathematical model.

Life Cycle Cost

Life cycle cost is the sum of all cost factors over the expected life of the production machinery. Appropriate discount factors may be applied for present value analysis. Total life cycle cost is derived as follows:

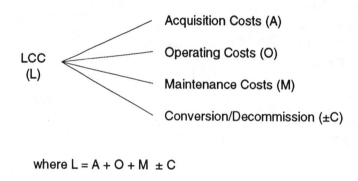

where L = A + O + M ± C

Breakdown of the values A, O, M and C are discussed in the following paragraphs.

Acquisition Costs

Acquisition costs are determined as follows:

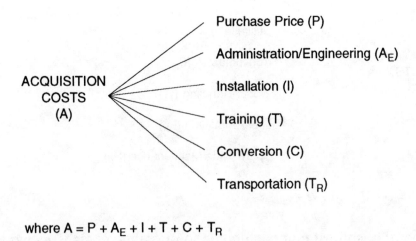

$$\text{where } A = P + A_E + I + T + C + T_R$$

A further description of each cost that contributes to acquisition costs is included below.

- Purchase price: The delivered price of the production machinery and equipment excluding transportation costs. This purchase price should account for currency differences and cost of money associated with payment schedules.

- Administration and engineering costs: Personnel, travel and runoff costs.

- Installation costs: The costs uniquely associated with the installation of the production machinery and equipment.

- Training costs: The costs associated with training of the work force to operate or maintain the particular production machinery and equipment.

- Conversion costs: The costs associated with conversion of the production machinery and equipment to a new product during its expected life.

- Transportation costs: The costs associated with the movement of production machinery and equipment from manufacturing location to user location.

Operating Costs

Operating costs are determined as follows:

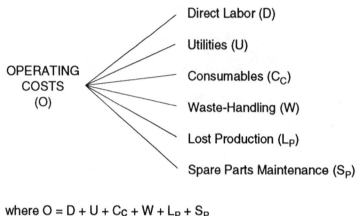

OPERATING
COSTS
(O)

- Direct Labor (D)
- Utilities (U)
- Consumables (C_C)
- Waste-Handling (W)
- Lost Production (L_P)
- Spare Parts Maintenance (S_P)

where $O = D + U + C_C + W + L_P + S_P$

A further description of each cost that contributes to operating costs is included below.

- Direct labor costs (operating): The total cost of direct labor to operate the production machinery and equipment over its expected life.

- Utilities costs: The total utilities consumption cost over the expected life of the production machinery and equipment. Includes air, steam, electricity, gas and water. Utilities costs can vary substantially and should be considered if utility resources are scarce.

- Consumable costs: The costs of all consumable items used by the production machinery and equipment over the expected life. Cost items to consider are coolant, filter media, etc.

- Waste-handling costs: The costs of collecting and disposing of waste products associated with the equipment.

- Cost of lost production: Lost production is due to machine failure and appropriate downtime costing must be agreed upon.

- Cost of maintaining spare parts: The cost of carrying and maintaining spare parts inventories to support the machinery and equipment.

Maintenance Costs

Maintenance costs are determined as follows:

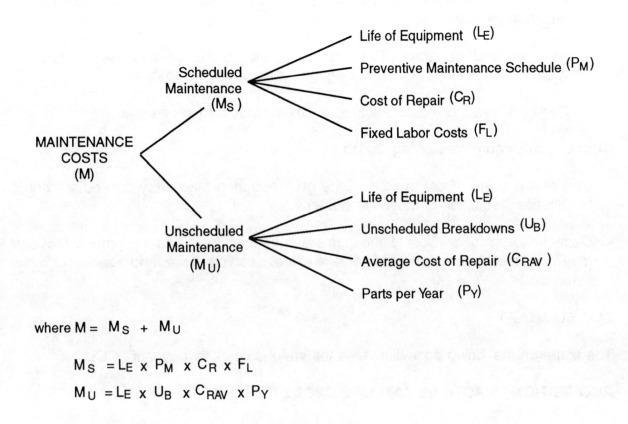

where $M = M_S + M_U$

$$M_S = L_E \times P_M \times C_R \times F_L$$

$$M_U = L_E \times U_B \times C_{RAV} \times P_Y$$

Scheduled and unscheduled maintenance costs are described below.

- Scheduled maintenance costs: Costs of material and labor associated with the preventive maintenance schedule during the expected life.

  - Life of equipment: See expected life in glossary.

  - Preventive maintenance schedule: The schedule of recurring maintenance actions to ensure long life of the equipment. The maintenance actions and frequency of occurrence are provided by the machinery and equipment supplier.

  - Cost of repair: Cost of labor and parts necessary to rectify the equipment from a failed state.

  - Fixed labor cost: Cost associated with maintaining a pool of skilled labor to service unscheduled breakdowns.

- Unscheduled maintenance costs: Cost of material and labor associated with unscheduled breakdowns during the expected life.

  - Unscheduled breakdowns: Resulting from equipment failure. See definition of failure in glossary.

  - Average cost of repair: The cost of repair actions caused by unscheduled breakdowns where the cost is averaged over several breakdowns.

  - Parts per year: The cost of all repair parts consumed in one year.

Conversion/Decommissioning Costs

- Conversion costs: Cost to convert the manufacturing machinery and equipment to handle the production of other components.

- Decommission costs: Cost to decommission the machinery and equipment. This cost may include such items as scrap/salvage value, cleaning of site, and disposal of waste by-products.

LCC SUMMARY

The figure on the facing page illustrates the various costs that impact LCC.

ACQUISITION PRACTICES AND LIFE CYCLE COSTING

Application of the LCC concept requires that acquisition of machinery and equipment be viewed from the total systems perspective. The common purchasing practice of considering only initial purchase price must be expanded to encompass the total life cycle cost elements for the system under consideration.

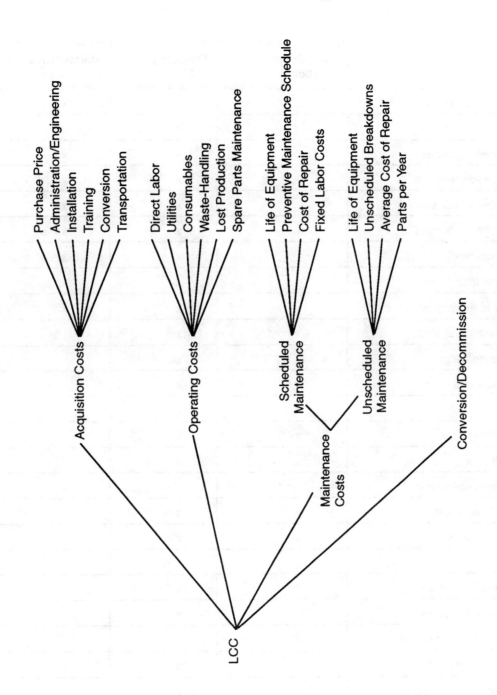

KEY LIFE CYCLE FACTORS

The table below is used in calculating acquisition, operating, maintenance and conversion/decommissioning costs.

	Acquisition + Costs	Operating + Costs	Maintenance +/- Costs	Conv/ Decomm.
1. Total hours worked per week		x	x	
2. Utilization Goal		x		
3. Mean time between failures		x	x	
4. Mean time between repairs		x	x	
5. Mean time between adjustments		x	x	
6. Mean time between changes		x	x	
7. Mean time to change		x	x	
8. Scheduled downtime		x	x	
9. Unit yield due to equipment	x	x	x	
10. Technician hourly wage rate		x		
11. Machine/operator ratio		x		
12. Operator hourly rate		x		
13. Tooling costs	x	x		
14. Electricity KW/R Rating	x	x		
15. Cost per KW	x	x		
16. Indirect material dollar/week			x	
17. Machine footprint (sq. ft.)	x	x		
18. Space valuation/sq. ft.	x	x		
19. Rework volume	x	x		
20. Materials costs/unit for rework		x		
21. Labor costs/unit for rework		x		
22. Equipment costs/unit of rework	x			
23. Subcontracting costs/unit for rework		x		
24. Handling costs/unit for rework		x		
25. Conversion/Decommissioning	x	x	x	

APPENDIX D

TRACKING & FEEDBACK SYSTEM FOR COMPONENT FAILURES

This appendix suggests a tracking and feedback system for component failures. If a scheme such as this is planned, it should be mutually agreed upon between supplier and user as to what data and parts should be returned. Although there are many differences between warranty and non-warranty actions, defective parts should be faithfully tagged regardless of whether or not the warranty is in force. It is very important that the system be rigidly adhered to if implemented. The accurate completion of meaningful data is critical to success of the system.

Figure D-1 provides a pictorial representation of a tracking and feedback system. The numbers in the figures are keyed to the following process:

1. A universal tag (see Figure D-2) is attached to the failed component with a copy to be entered in the plant data base.
2. Data is entered in the plant data base.
3. A copy of the universal tag is sent to the manufacturing machinery and equipment supplier.
4. The component is sent to the component manufacturer.
5. An evaluation of the component is made to determine root cause of failure.
6. An identification of required corrective action is made.
7. A report is generated by the manufacturer and sent to the user and supplier.
8. The report is entered in the plant and/or supplier data base.
9. A report is sent from the plant data base to the user on the plant floor with all current information and reconciled activity.
10. An exception report is generated from the data base to indicate recurring failures. The manufacturer/user/supplier team takes action.

The action plan to address recurring failures should involve the manufacturing machinery and equipment supplier and the user. This team should resolve issues concerning the design, manufacture and the application of the part.

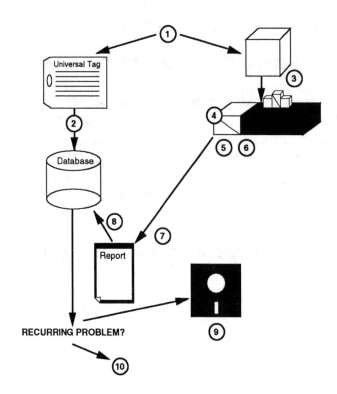

Figure D-1. Tracking and Feedback

Figure D-2. Universal Tag

FEEDBACK DATA REQUIREMENTS

The following is the recommended <u>minimum</u> data for each failure/downtime incident:

- Plant/Location #
- Fault code
- Initial observation
- Time of event
- Time event cleared
- Response time
- Repair time
- Duration of downtime
- Total on-time (Life)
- Total run-time (Life)
- Operation #
- Station #
- Brass tag #
- Component failure: part #
- Repair comments

In addition to the above minimum requirements, the following data are <u>highly desirable</u> and allows a more comprehensive data base to be formed:

- Description of root cause (if known)
- Machine state (such as proximity switch status, slide position, Input/Output data, etc.)
- Yield
- Performance efficiency

ACCESS TO DATABASE BY SUPPLIERS

The following paragraphs suggest several methods of providing R&M data to the supplier. These data should be available in electronic form, preferably in ASCII text with a specified delimiter or on a standard commercially available data base.

Periodic Supplier Plant Visits

The user plant should allow access to operational data by the supplier. Security issues have to be addressed to ensure this accessibility. The supplier should take the initiative in collecting these data by periodic visits to the plant and analysis of the data for set periods of time.

Local (Station) Data Dumps

Event data should be captured and stored at the local level. Capture and storage of event data at the station or zone level makes it unnecessary for the supplier to access the plant monitoring data base. The supplier is then allowed access to event data specific to his equipment anytime supplier personnel are on-site.

Remote Access

Plants should consider the ability of the equipment to store event data at the local or zone level for remote data access. The supplier could access these data at will via remote dial-up through a built in communication link to the machine controller.

Delay Studies

To improve the baseline reliability of specific lines, plant operations should encourage supplier visits and provide suppliers open access to the plant to conduct delay studies. These delay studies should be joint ventures with supplier and plant personnel.

BENEFITS OF PLANT MONITORING SYSTEMS AND R&M DATA IN PLANT OPERATIONS

Increased Uptime

Critical to increased competitiveness is the ability to increase throughput with minimum cost increase. For this reason, plant management is very interested in getting improved uptime of equipment. R&M data are important aids to improve uptime.

Rapid Bottleneck Identification

Production monitoring systems should have the ability to pin-point production bottlenecks quickly. Plant management can then focus the plant manufacturing engineering resources on root cause analysis.

Improved Root Cause Analysis & Failure Analysis

R&M data provide a basis for root cause analysis. Root cause analysis is needed for continuous improvement and improved equipment designs. Appendix A provides more information concerning Root Cause Analysis and Failure Analysis.

Continuous Improvement (CI) Monitoring - Reliability Growth

Data feedback is essential for any continuous improvement program. It allows CI teams to focus on the critical areas, and provides quantifiable feedback of the impact of continuous improvement. The continuous improvement activity is dual. It occurs within the user plant but also in the supplier organizations. The supplier must have access to the data to drive the continuous improvement activity.

Reduced Life Cycle Cost (LCC)

R&M data provide a basis for reducing operating expenses associated with the LCC by providing the information necessary to help identify problem areas.

Improved Production Equipment for Future Programs

Continuous feedback from several user plants provides the supplier with vital information for improved designs for future programs. The data feedback can be used to update simulation models with real data for concept and design stages of new systems.

APPENDIX E

SOURCES OF R&M DATA FROM THE USER PLANT

This table lists the sources from which R&M data may be obtained. Also included in the table are the traditional and the desired methods of obtaining the data.

R&M Data Sources

SOURCES OF DATA	TRADITIONAL	DESIRED
Central maintenance records: -Dispatch system work order -Repair card system -Universal tag system -PM records -Supplier service records -Machine history	**Plant Monitoring & Maintenance Feedback System:** **Plant with Existing Machine Tools** -typically 100% written manual/paper records -provides inadequate, untimely, maintenance data **Plant Monitoring & Maintenance Feedback System: New plant & new machine tools** -automated, but rarely being used as intended -maintenance history by paper records, some use of electronic data base systems but not linked to the plant monitoring system	Data entered or captured and stored in the machine Tool Controller. -available for remote access upon demand -maintenance history (corrective feedback) data available and frequently used -maintenance database systems designed to provide root cause analysis and summary data (made useful and easy to use)
Quality Control Charts (Predictor and Diagnostic tools for R&M)	Post or in-process gaging stores SPC data and provides screen presentation of chart data	SPC software has expert system to suggest possible root causes and/or assist in root cause analysis
Special Machine Tool Studies: -Delay Studies -Laser Alignment -High Speed Photography/Video Signature Analysis: -Vibration -Noise -Force/torque -Power consumption -Chemical analysis -Thermography	Supplier or user performs independently	Joint participation by supplier-user teams
Inventory Systems: -Spare parts consumption -Consumables (inserts,coolant, oils)	Inventory system not easily accessible to maintenance personnel	Computer inventory system accessible from the maintenance system
Equipment Monitor (excludes CNC) & Cell Controllers	PLC - limited failure event history captured	PC based - failure event history captured and stored for remote access. Auto cycle not allowed to resume until failure is documented either via control console or paper operator log/failure report.

GLOSSARY

Accelerated Life Testing
Testing to verify design reliability of machinery/equipment much sooner than if operating typically. This is intended especially for new technology, design changes, and ongoing development.[1]

Acceptance Test [Qualification Test]
A test to determine machinery/equipment conformance to the qualification requirements in its equipment specifications.

Accessibility
The amount of working space available around a component sufficient to diagnose, troubleshoot, and complete maintenance activities safely and effectively. Provision must be made for movement of necessary tools and equipment with consideration for human ergonomic limitations.

Actual Machine Cycle Time (Process Cycle Time)
Actual time to process a part or complete an operation. Specifically, the shortest period of time at the end of which a series of events in an operation is repeated.

Allocation
The process by which a top-level quantitative requirement is assigned to lower hardware items/subsystems in relation to system-level reliability and maintainability goals.[2]

Availability
A measure of the degree to which machinery/equipment is in an operable and committable state at any point in time. Specifically, the percent of time that machinery/equipment will be operable when needed.

Built-in-Test (BIT)
The self-test hardware and software that is internal to a unit to test the unit.[3]

Built-in-Test Equipment (BITE)
A unit which is part of a system and is used for the express purpose of testing the system. BITE is an identifiable unit of a system.[3]

Corrective [Unscheduled, Unplanned, Repair] Maintenance
All actions performed as a result of failure, to restore a machine to a specified condition. Corrective maintenance can include any or all of the following steps: localization, isolation, disassembly, interchange, reassembly, alignment, and checkouts.

Delay Study

A continuous study over an extended period of time (say 2 weeks) where every incidence of downtime is recorded along with the apparent cause.

Design Machine Cycle Time [Process Cycle Time]

Specified time to process a part or complete an operation. Specifically, the shortest period of time at the end of which a series of events in an operation is repeated.

Durability Life [Expected Life]

A measure of useful life, defining the number of operating hours (or cycles) until overhaul is expected or required.[5]

Failure

An event when machinery/equipment is not available to produce parts at specified conditions when scheduled or is not capable of producing parts or perform scheduled operations to specification. For every failure, an action is required.[7]

Failure Analysis (FA)

The logical systematic examination of a failed item, its construction, application and documentation to verify the reported failure, identify the failure mode and determine the failure mechanism and its basic failure cause. To be adequate, the procedure must determine whether corrective action is warranted and if so, provide information to initiate corrective action.[1]

Failure Effect

The consequence of the failure.[1]

Failure Mode

The manner by which a failure is observed. Generally a failure mode describes the way the failure occurs and its impact on equipment operation.[13]

Failure Rate

Number of failures per unit of gross operating period in terms of time, events, cycles, or number of parts.

Failure Mode Analysis (FMA)

For each critical parameter of a system, determining which malfunction symptoms appear just before, or immediately after, failure.[1]

Failure Mode and Effects Analysis (FMEA)

A technique to identify each potential failure mode and its effect on machinery performance (see Appendix A).

Fault Tree Analysis (FTA)

A top-down approach to failure analysis starting with an undesirable event and determining all the ways it can happen.

Gross Operating Time

Total time that the machine is powered and producing parts.[7]
Gross operating time = Net operating time + Scrap time

Infant Mortality

Early failures that exist until debugging eliminates faulty components, improper assemblies and other user and manufacturer learning problems, and until the failure rate lowers.[1]

Life Cycle

The sequence of phases through which machinery/equipment passes from conception through decommission.[4]

Maintainability

A characteristic of design, installation and operation, usually expressed as the probability that a machine can be retained in, or restored to, specified operable condition within a specified interval of time when maintenance is performed in accordance with prescribed procedures.

Maintenance

Work performed to maintain machinery and equipment in its original operating condition to the extent possible; includes scheduled and unscheduled maintenance but does not include minor construction or change work.[7]

Mean Cycle Between Failures (MCBF)

The average cycles between failure occurrences. The sum of the operating cycles of a machine divided by the total number of failures.[8]

Mean Cycle-To-Repair (MCTR)

The average cycles to restore machinery or equipment to specified conditions.[8]

Mean Time Between Failures (MTBF)

The average time between failure occurrences. The sum of the operating time of a machine divided by the total number of failures.[8]

___

Mean Time-To-Repair (MTTR)

The average time to restore machinery or equipment to specified conditions.[8]

Net Operating Time

Total time that the machine is producing parts, as a first pass, to specifications.

Non-operating Time

Total time that the machinery/equipment is up but not running due to blocking, starvation and/or administrative time.

Overall Equipment Effectiveness (OEE)

The product of three measurements:

Percentage of time the machinery is available (**Availability**) x how fast the machinery or equipment is running relative to its design cycle (**Performance efficiency**) x percentage of the resulting product within quality specifications (**Yield**).

The overall machine effectiveness for the machinery or equipment is calculated by:

OEE = Availability x Performance efficiency x Yield[9]

Overhaul

A comprehensive inspection and restoration of machinery/equipment, or one of its major parts, to an acceptable condition at a durability time or usage limit.[1]

Predictive & Preventive [Scheduled, Planned] Maintenance

All actions performed in an attempt to retain a machine in specified condition by providing systematic inspection, detection, and prevention of incipient failures.[4]

Quality Function Deployment (QFD)

A discipline for product planning and development or redesigning an existing product in which key user wants and needs are deployed throughout an organization. QFD provides a structure for ensuring that users' wants and needs are carefully heard, then directly translated into a company's internal technical requirements from component design through final assembly.

Redundancy

The existence of more than one means for accomplishing a given function. Each means of accomplishing the function need not necessarily be identical.[10]

Reliability

The probability that machinery/equipment can perform continuously, without failure, for a specified interval of time when operating under stated conditions.[11]

Reliability Growth

Machine reliability improvement as a result of identifying and eliminating machinery or equipment failure causes during machine testing and operation.

Scheduled [Planned] Downtime

The elapsed time that the machine is down for scheduled maintenance or turned off for other reasons.[7]

Testability

A design characteristic allowing the following to be determined with a given confidence and in specified time: location of any faults, whether an item is inoperable, operable but degraded, and/or operable.[1]

Time to Repair (TTR)

Total clock time from the occurrence of failure of a component or system to the time when the component or system is restored to service (i.e. capable of producing good parts or performing operations within acceptable limits). Typical elements of repair time are: diagnostic time, troubleshooting time, waiting time for spare parts, replacement/fixing of broken parts, testing time, and restoring.[12]

Total Downtime

The elapsed time during which a machine is not capable of operating to specifications.[8]

Total Downtime = Scheduled Downtime + Unscheduled Downtime

(Also see time diagram on glossary page 6).

Unscheduled [Unplanned] Downtime

The elapsed time that the machine is incapable of operating to specifications because of unanticipated breakdowns.[7]

Uptime

Total time that a machine is on-line (powered up) and capable of producing parts.[8]

Uptime = Gross operating time + Non-operating time

(Also see time diagram on glossary page 6).

Yield

The fraction of products meeting quality standards produced by the machinery or equipment.[9]

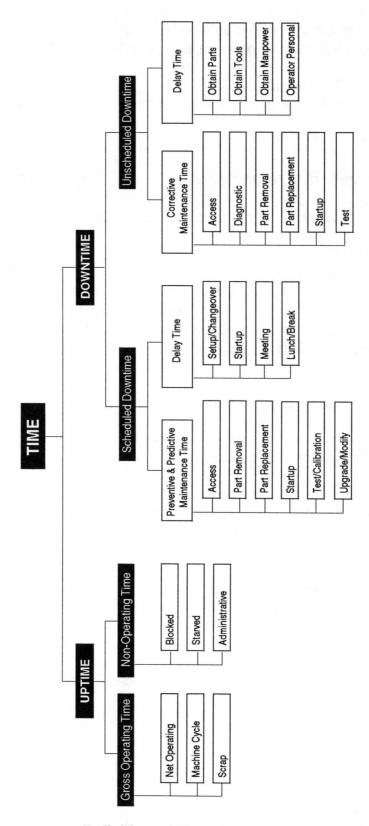

Definition of Time Elements

Term Clusters

The term clusters help users locate different expressions that correspond to a main term. The previous pages of this Glossary contain the definitions of each term as it pertains to this guideline.

FAILURE
Failure
Failure Analysis (FA)
Failure Mode
Failure Rate
Failure Mode Analysis (FMA)
Failure Mode and Effects Analysis (FMEA)
Fault Tree Analysis (FTA)
Infant Mortality

LIFE
Durability Life [Expected Life]
Gross Operating Time

MAINTENANCE
Corrective [Unscheduled, Unplanned, Repair] Maintenance
Maintainability
Maintenance
Predictive & Preventive [Scheduled, Planned] Maintenance

MISCELLANEOUS
Accessibility
Allocation
Delay Study
Life Cycle
Overall Equipment Effectiveness (OEE)
Overhaul
Yield

RELIABILITY
Availability
Redundancy
Reliability
Reliability Growth

TEST

Accelerated Life Testing
Acceptance Test [Qualification Test]
Built-in-Test (BIT)
Built-in-Test Equipment (BITE)
Testability

TIME

Actual Machine Cycle Time (Process Cycle Time)
Design Machine Cycle Time [Process Cycle Time]
Mean Cycle Between Failures (MCBF)
Mean Cycle-To-Repair (MCTR)
Mean Time Between Failures (MTBF)
Mean Time-To-Repair (MTTR)
Net Operating Time
Non-operating Time
Scheduled [Planned] Downtime
Time to Repair (TTR)
Total Downtime
Unscheduled [Unplanned] Downtime
Uptime
(Also see time diagram on glossary page 6).

Glossary References

1. Omdahl, T.P. *Reliability, Availability, and Maintainability (RAM) Dictionary,* ASQC Quality Press, 1988.

2. MIL-STD-785, *Reliability Program for Systems and Equipment Development and Production.*

3. MIL-STD-2084(AS), *General Requirements for Maintainability of Avionic and Electronic Systems and Equipment.*

4. MIL-STD-721C, *Definitions of Terms for Reliability and Maintainability.*

5. Society of Automotive Engineers, Inc. *Glossary of Reliability Terminology Associated with Automotive Electronics,* SAE J1213/2, October 1988.

6. ANSI/IEEE STD 500-1984.

7. General Motors Corporation. *GM Reliability and Maintainability Guideline for Machinery and Equipment,* October 1991.

8. ANSI/IEEE STD C37.1, 1987.

9. Nakajima, Seiichi. *Introduction to TPM, Total Productive Maintenance,* Productivity Press, Norwalk, CT, 1990.

10. MIL-STD-721B.

11. Institute of Electrical and Electronic Engineers, Inc. *IEEE Recommended Practice for Design of Reliable Industrial and Commercial Power Systems,* Wiley-Interscience, New York, NY, 1980.

12. ANSI/IEEE STD 493, 1980.

13. MIL-STD-1629A.

BIBLIOGRAPHY

A number of publications are available on the subject of R&M. Some frequently cited references include:

1.   Arsenault, J. E. and J. A. Roberts. *Reliability and Maintainability of Electronic Systems,* Computer Science Press, Rockville, MD, 1980.

2.   Blanchard, B.S. *Logistics Engineering and Management,* 2nd Ed., Prentice-Hall, Inc., Englewood Cliffs, NJ, 1981.

3.   Bompas-Smith, J.H. *Mechanical Survival: The Use of Reliability Data,* McGraw-Hill, Inc., New York, NY, 1973.

4.   Codier, E. O. "Reliability Growth in Real Life," *Proceedings, 1968 Annual Symposium on Reliability,* New York: IEEE, January 1968.

5.   Crow, L. H. "On Tracking Reliability Growth," *Proceedings 1975 Annual Reliability and Maintainability Symposium,* New York: IEEE, January 1975.

6.   Crow, L. H. "Reliability Growth Potential from Delayed Fixes," *Proceedings 1983 Annual Reliability and Maintainability Symposium,* New York: IEEE, January 1983.

7.   Crow, L. H. "Reliability Management, Reliability Growth Strategies, Growth Potential Estimation, Confidence Bounds," *Proceedings 1984 Annual Reliability and Maintainability Symposium,* New York: IEEE, January 1984.

8.   Duane, J.T. "Learning Curve Approach to Reliability Monitoring," *IEEE Transaction on Aerospace,* Vol. 2, 1964.

9.   Ford Motor Company. *Reliability and Maintainability Guideline, The R&M Approach to Competitiveness,* January 1990.

10. Henley, E.J. and H. Kumamoto. *Reliability Engineering and Risk Assessment,* Prentice-Hall, Inc., Englewood Cliffs, NJ, 1981. Also published as *Probabilistic Risk Assessment,* IEEE Press, New York, NY.

11. Institute of Electrical and Electronics Engineers, Inc. *IEEE Recommended Practice for Design of Reliable Industrial and Commercial Power Systems,* Wiley-Interscience, New York, NY, 1980.

12. Institute of Electrical and Electronics Engineers, Inc. *IEEE Standard Dictionary of Electrical and Electronics Terms,* Fourth Edition, IEEE, New York, 1988.

13. Ireson, W.G. (editor), *Handbook of Reliability Engineering & Management,* McGraw-Hill, Inc., New York, NY, 1966.

14. Kapur, K. C. and L. R. Lamberson. *Reliability in Engineering Design,* John Wiley & Sons, Inc., New York, 1977.

15. General Motors Corporation. *GM Reliability and Maintainability Guideline for Machinery and Equipment,* October 1991.

16. Kececioglu, D. *Reliability Engineering Handbook,* Vol. I, II, Prentice-Hall, Inc., Englewood Cliffs, NJ, 1991.

17. MIL-HDBK-338, *Electronic Reliability Handbook.*

18. MIL-HDBK-5F, *Metallic Materials and Elements for Aerospace Vehicle Structures.*

19. Nakajima, Seiichi. *Introduction to TPM, Total Productive Maintenance,* Productivity Press, Norwalk, CT, 1990.

20. O'Conner, P.D.T. *Practical Reliability Engineering,* 3rd Ed., Heyden & Sons, Inc., Philadelphia, PA, 1981.

21. Raheja, Dev G. *Assurance Technologies, Principles and Practices,* McGraw-Hill, Inc., New York, NY, 1991.

22. Selby, J. D. and S. G. Miller. "Reliability Planning and Management-RPM," *Symposium for Reliability and Maintainability Technology for Mechanical Systems,* Washington: ADA, April 1972.

23. Society of Automotive Engineers, Inc. *Glossary of Reliability Terminology Associated with Automotive Electronics,* SAE J1213/2, October, 1988.

24. Seldon, M.R. *Life Cycle Costing: A Better Method of Government Procurement,* Westview Press, Boulder, CO, 1979.

25. Von Alven, W.H. Ed. *Reliability Engineering,* Prentice-Hall, Inc., Englewood Cliffs, NJ, 1964.

26. Block, Heinz P. and Geitner, Fred K. *An Introduction to Machinery Reliability Assessment,* Van Nostrand Reinhold, New York, NY, 1990